c   problem lay in the measurements
of cosmic distances based on the notion of 'standard candles'. These
distances were substantially underestimated, inflating the value of the
Hubble constant derived from them. This was a troubled history. It took
more than 70 years to sort it out.

Theorists found it difficult to understand how galaxies might have
formed without invoking the existence of a wholly unknown form of
matter. This had to be much more abundant in the universe than the
more familiar matter in stars and interstellar gas and dust. They called it
dark matter. The way that both dark matter and ordinary visible matter are today scattered across the cosmos—in a web of great clumps
and voids—is explained by quantum fluctuations, impressed upon the
primordial universe by cosmic inflation, an insane burst of exponential
expansion thought to have occurred within a hundredth of a millionth
of a trillionth of a trillionth of a second after the Big Bang. In the late
1990s, it was discovered that, against all expectations, the rate of expansion of the universe is actually accelerating. This can be explained by
proposing that 'empty' space itself carries energy which acts like a kind of
anti-gravity, carrying matter apart just as gravity tries to draw it together.
This is called dark energy.

These concepts have been combined with the results of exquisitely
detailed satellite studies of the cosmic background radiation, surveys of
the large-scale structure of the universe, and painstaking measurements
of cosmic distances to give us a theory, known variously as the standard
model of inflationary Big Bang cosmology, or the Λ-CDM (lambda,
cold dark matter) model. It has many parameters, but the 'concordance'
achieved by drawing on data from different sources gives confidence in
their determination. Once it was off to a good start, we now think we
understand how our universe got to here, with the properties we find it
to have today.

On the one hand, this theory is a triumph of modern science and the
human intellect. On the other hand, the theory is deeply unsatisfactory.
There are nagging doubts about cosmic inflation. And we have no real

idea what dark matter or dark energy are, which means that we cannot properly explain about 95% of the universe.

So far, so familiar. But the full story of the development of observational astronomy, astrophysics, and physical cosmology over the past 125 years is told less often, and so is less familiar. *Discordance* traces the history of cosmic distance measurements and the determination of the Hubble constant, from the turn of the twentieth century to the present. As we might expect, as new observations have continually refined the lower rungs of the cosmic 'distance ladder' used by astronomers, so determinations of the Hubble constant have become more and more precise. These are *measurements*, based directly on empirical observations. They are also 'late-universe' measurements: they are limited to objects that appeared more recently in the history of the universe. They do not require the presumption of a specific cosmological model.

Likewise, cosmological parameters derived from studies of the cosmic background radiation and the large-scale disposition of galaxies have also become more precise. These studies have involved a succession of extraordinary ground-based, balloon-borne, and space-based missions. Some of these parameters are derived directly by fitting the data gathered by the instruments. Others, including the Hubble constant, are derived more indirectly. These are 'early-universe' *predictions*. No parameters can be derived without first presuming a specific cosmological model.

If the Λ-CDM model is correct and all the observations are being done properly, then the model-dependent, early-universe predictions and the model-independent, late-universe measurements of the Hubble constant should agree.

Here's where things get interesting. When we compare the most recent results from late-universe distance ladder measurements using certain types of standard candles with the most recent early-universe predictions, they do not quite agree. The difference is small—about 7%—but large enough to be significant given the precision of the individual results. This is the *Hubble tension*. Some late-universe measurements

DISCORDANCE

THE TROUBLED HISTORY OF THE HUBBLE CONSTANT

JIM BAGGOTT

OXFORD
UNIVERSITY PRESS

OXFORD
UNIVERSITY PRESS

Great Clarendon Street, Oxford, OX2 6DP,
United Kingdom

Oxford University Press is a department of the University of Oxford.
It furthers the University's objective of excellence in research, scholarship,
and education by publishing worldwide. Oxford is a registered trade mark of
Oxford University Press in the UK and in certain other countries.

© Jim Baggott 2025

The moral rights of the author have been asserted.

All rights reserved. No part of this publication may be reproduced, stored in a retrieval system,
transmitted, used for text and data mining, or used for training artificial intelligence, in any form or
by any means, without the prior permission in writing of Oxford University Press, or as expressly
permitted by law, by licence or under terms agreed with the appropriate reprographics rights
organization. Enquiries concerning reproduction outside the scope of the above should be sent
to the Rights Department, Oxford University Press, at the address above.

You must not circulate this work in any other form
and you must impose this same condition on any acquirer.

Published in the United States of America by Oxford University Press
198 Madison Avenue, New York, NY 10016, United States of America

British Library Cataloguing in Publication Data
Data available

Library of Congress Control Number: 2025936248

ISBN 9780192864062

DOI: 10.1093/oso/9780192864062.001.0001

Printed and bound in the UK by
Clays Ltd, Elcograf S.p.A.

The manufacturer's authorised representative in the EU for product safety is
Oxford University Press España S.A., Parque Empresarial San Fernando de Henares,
Avenida de Castilla, 2 – 28830 Madrid (www.oup.es/en or product.safety@oup.com).
OUP España S.A. also acts as importer into Spain of products made by the manufacturer.

Links to third party websites are provided by Oxford in good faith and
for information only. Oxford disclaims any responsibility for the materials
contained in any third party website referenced in this work.

MIX
Paper | Supporting
responsible forestry
FSC® C018072

For John Heilbron

Preface

It's a familiar story, and most readers of popular science will know broadly how it goes. According to the Big Bang theory, the universe 'began' about 13.8 billion years ago in a fiery origin of space and time, and of matter and radiation. Some 380,000 years later, the universe had expanded and cooled sufficiently to allow free protons and bare helium nuclei to combine with free electrons. Radiation that had danced unchecked among the free electrical charges in the mother of all electrical storms now had nowhere to go. It was released, flooding the universe with light, some of it visible, though of course there was nobody around to see it. The first light faded as space further expanded and stretched, disappearing everywhere into the cosmic background. Light returned only when the first stars and galaxies appeared.

Although a few theorists, such as Georges Lemaître, had anticipated the possibility that our universe might be expanding, it was astronomers Vesto Slipher, Edwin Hubble, and Milton Humason who discovered the expanding universe in 1929. The further away a galaxy lies, the faster it appears to be moving away from us, in a simple linear relationship governed by the *Hubble constant*. Galaxies do move through space, under the influence of the gravitational pull of their nearest neighbours, at speeds typically measured in hundreds of kilometres per second, or km/s. But the recession speeds of galaxies have a different origin. They are being carried away by the expansion of the space that lies between them, at speeds ranging from many thousands of km/s to a substantial fraction of the speed of light.

But things didn't quite add up. Early estimates of the value of the Hubble constant implied that the universe must be *younger* than many made using different standard candles are not entirely consistent, for reasons which are presently unclear, but all tend to be a little higher than the early-universe predictions.

If the Hubble tension is real (and we should know either way in a few years), there aren't too many options. Either the distance ladder measurements or the measurements of the cosmic background suffer from some systematic error that has so far been overlooked. Or the Λ-CDM model must be modified or extended in some way that involves 'new physics'. Some argue that the time has come to acknowledge the failings of the model and rethink cosmology completely.

This is science perched right at the frontier. Our scientific answers can sometimes be of only fleeting interest, for it is often the *questions* that fascinate. As *Discordance* will recount, in the troubled history of the Hubble constant and the sometimes uneasy relationship between cosmological theory and observation, we have never been short of fascinating questions.

Contents

CONTENTS

About the Author

Jim Baggott is a freelance science writer. He gained a B.Sc. (1978) in chemistry at the University of Manchester and completed a D.Phil. (1981) at Oxford University. He worked as a postgraduate research fellow at Oxford and at Stanford University in California. He returned to England to take up a lectureship in chemistry at the University of Reading. After five years of academic life, he decided on a complete change of career and worked in the oil industry for 11 years before setting up his own independent business consultancy and training practice.

He maintains a broad interest in science, philosophy, and history, and writes on these subjects in what spare time he can find. He was awarded the Marlow Medal by the Royal Society of Chemistry in 1989 and a Glaxo Science Writer's prize in 1992. His book *Mass: The Quest to Understand Matter from Greek Atoms to Quantum Fields*, won the Cosmos Prize for Science Writing in 2020.

Prologue

Starry Messengers

I've never been very good at identifying constellations. Some folks have a knack for pointing upwards at the sky in seemingly random directions on a clear night and smugly declaring the names of these obscure patterns in the starry points of light. Not me. The best I can manage is the Plough (or Big Dipper in the US) and Orion. The former has a distinctive pattern of seven stars, four of which form the bucket or bowl, and three of which form the handle. The latter is readily identified by the three stars which form the belt, from which it's quite straightforward to locate the shoulders and the skirt. It seems there's a head, shield, club, and sword somewhere but in our urban, light-polluted night sky these details are beyond my competence and imagination.

Orion's Belt is also known variously as the Three Kings, Three Marys, or Three Sisters. It is comprised of (from left to right, in the northern sky) Alnitak (Zeta Orionis), Alnilam (Epsilon Orionis), and Mintaka (Delta Orionis).[1] This is where we learn that, when it comes to observations of the night sky made with the naked eye, appearances can be deceptive. Alnitak is actually three stars, with blue supergiant and blue dwarf stars orbiting each other in a binary system which is in turn orbited by a third star every 1500 years. The blue dwarf star was only discovered in 1998. Alnilam, the Belt's middle star, is a blue supergiant. The main component of Mintaka is an eclipsing binary star system, but there are another two stars close by that may be companions.

And so we come to the main lesson. The three stars that in human culture may have been associated with Orion's Belt for at least 32,000 years[2] share the same origin and have moved through space together, but their alignment across the night sky is an illusion. The stars appear to have similar visual brightness, or what astronomers call *apparent magnitude*, but they all lie at different distances from Earth. In terms of the star's intrinsic brightness or *absolute magnitude*, the middle star Alnilam is the brightest of the three.[3] It is also the furthest away.

To decipher the messages encrypted in the light from distant stars, astronomers of the early twentieth century needed to untangle the complex relationships between the brightness and distances of objects in the night sky, to find a way to measure these distances, and to study the spectrum of the light absorbed or emitted by these objects' outer layers. Remarkably, these relationships would help to provide an answer to some of the oldest and deepest of the 'big questions' of human existence. Did the universe have a beginning? Or has it existed in its current form for all eternity?

The Brightness of Stars

I was born in 1957, and was a teenager when the British government finally bit the bullet and adopted a decimalized version of its currency, in 1971. Prior to this date, the British system of pounds, shillings, and pence (£sd) prevailed. There were 12 pennies (d) to a shilling (s), 20s to a pound (£), and so 240d to a £. This 12–20 system had been introduced by Charlemagne in the eighth century, to be abandoned first by America in 1792, some years after the Declaration of Independence, then by France in 1795 (following the Revolution), then by most of Continental Europe by the middle of the nineteenth century. Britain held out due likely to a mix of exceptionalism and stubbornness. But wariness of the problems and the costs associated with the process of decimalization were undoubtedly also factors.

To a certain extent, this kind of logic—constrained by a staunch conservatism—explains why astronomers continue with an even older and more archaic approach to the measurement of the relative brightness of stars, or stellar magnitude. The great Greek astronomer and mathematician Hipparchus of Nicaea is frequently credited with inventing the system of classifying the brightness of the visible stars. Alas, although this system was certainly available in the second century, at the time Ptolemy was writing the *Almagest*, there is no evidence that helps us to trace its origin to Hipparchus.

Whatever its origin, the system is simple and logical. Yet, as we will soon see, it does create some challenges for those unfamiliar with it. The brightest stars in the night sky were designated variously as 1st class, 1st magnitude, or magnitude-1, continuing through to the dimmest, designated as 6th magnitude or magnitude-6. These classifications were based on naked-eye observations, so the magnitudes in question are apparent magnitudes. On a clear night, Hipparchus (if indeed it was he) would have had the benefit of near-zero light pollution. In today's polluted night skies you'll be lucky to observe 4th magnitude stars. This is a great pity, as most of the objects that are in principle observable to the naked eye are of 5th magnitude and fainter.

Usually, 'bigger' or 'brighter' implies a larger measure. But this system works in reverse. The brightest stars are magnitude-1, the next brightest magnitude-2, and so on. So, the brighter the star, the *lower* its apparent magnitude. Now, the human eye is not the most reliable or reproducible of measuring instruments, and as the science of astronomy developed in the nineteenth century with the aid of telescopes, photography, and photometry, its reach extended substantially beyond the immediately 'visible' stars. By the mid-to-late nineteenth century, brightness was something that could be recorded, measured, and compared, rather than something that could only be judged by individual astronomers. By carefully timing the exposure of starlight on a photographic plate, and rigging the telescope to mechanically track a star in the night sky as the Earth turns on its axis, brightness could be translated into an image whose size could be measured. The larger the size of the spot recorded for a given exposure time, the brighter the star.

It became urgently necessary to replace the ancient system of crude magnitude classification with a continuous magnitude scale. If you think of the magnitude classes as individual fence posts—1, 2, 3, etc.—astronomers needed to erect the fence that would link them together and allow for magnitudes lying between the posts—1.7, 2.3, 3.9, and so on.

Astronomers understood that the system of magnitude classification was based on the workings of the human eye, which is a non-linear light detector in the sense that our perception of brightness doesn't scale in one-to-one correspondence with the apparent magnitudes of bright objects. Currency is a linear, additive scale—six shillings is six times more money than one shilling. In contrast, astronomers had found that a magnitude-1 star is one hundred times brighter than a magnitude-6 star. This suggests a multiplicative, rather than additive, scale. Magnitude-1 is a certain factor brighter than magnitude-2. Magnitude-2 is the same factor brighter than magnitude-3, and so on. The key question was: what is this factor?

This was, perhaps, the opportunity to abandon the old magnitude classification and adopt a scale based on factors of ten that works forwards. But no. Astronomers decided instead to shoehorn their observations into the old scheme. Some suggested different multiplicative factors, such as 2.43 and 2.83. In 1856, the English astronomer Norman Pogson suggested 2.512—now known as Pogson's ratio—for 'convenience of calculation'.[4] On this scale, magnitude-1 is 2.512 times brighter than magnitude-2, magnitude-2 is 2.512 times brighter than magnitude-3 (and so magnitude-1 is $(2.512)^2 = 6.310$ times brighter than magnitude-3), and so on. It's no coincidence that magnitude-1 is then $(2.512)^5 = 100$ times brighter than magnitude-6—see Fig. 1(a). The curvature in this graph illustrates the 'non-linear' nature of the relationship between apparent magnitude and relative brightness.

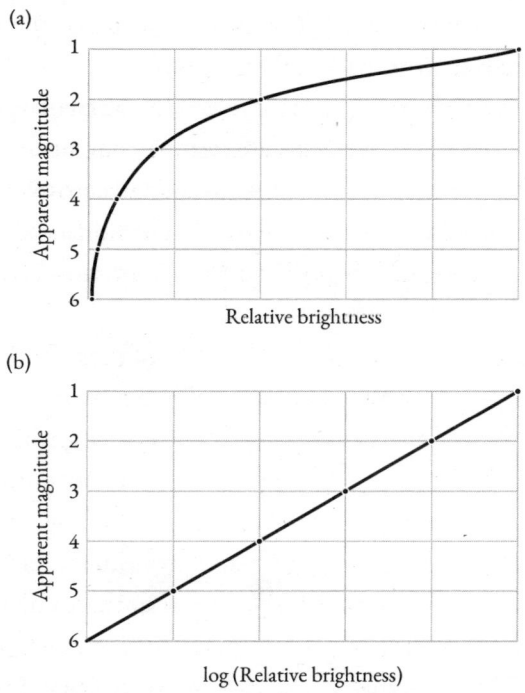

Fig. 1 (a) The apparent magnitudes of stars were originally organized into six classes in descending order of brightness, forming a multiplicative scale such that magnitude-2 is 2.512 times dimmer than magnitude-1, etc. (b) The scale can be 'linearized' by taking the logarithm of the brightness.

Whilst this is all very helpful, interpolating between the classes of magnitude can be tricky, and scientists much prefer to deal with linear relationships whenever they can. The scale can be made linear if the apparent magnitude is assumed to be related to the *logarithm* of the relative brightness, as shown in Fig. 1(b). At the time this was entirely consistent with developments in our understanding of human perception and psychology. Research published by the German experimental psychologist Gustav Fechner in 1860 supported a law—called the Weber–Fechner law—which relates the psychological intensity of human sensation (such as sound and vision) to the logarithm of the physical intensity of the stimulus.[5] We now know that human sensation is governed by a power law rather than a logarithmic law, but the logarithmic basis for the magnitude scale was retained.

We should note that this scale is entirely relative and we mustn't forget that it works in reverse. The brighter the star the lower its magnitude. The scale is most useful when we want to determine the apparent magnitude of one star relative to another. It turns out that Alnitak,

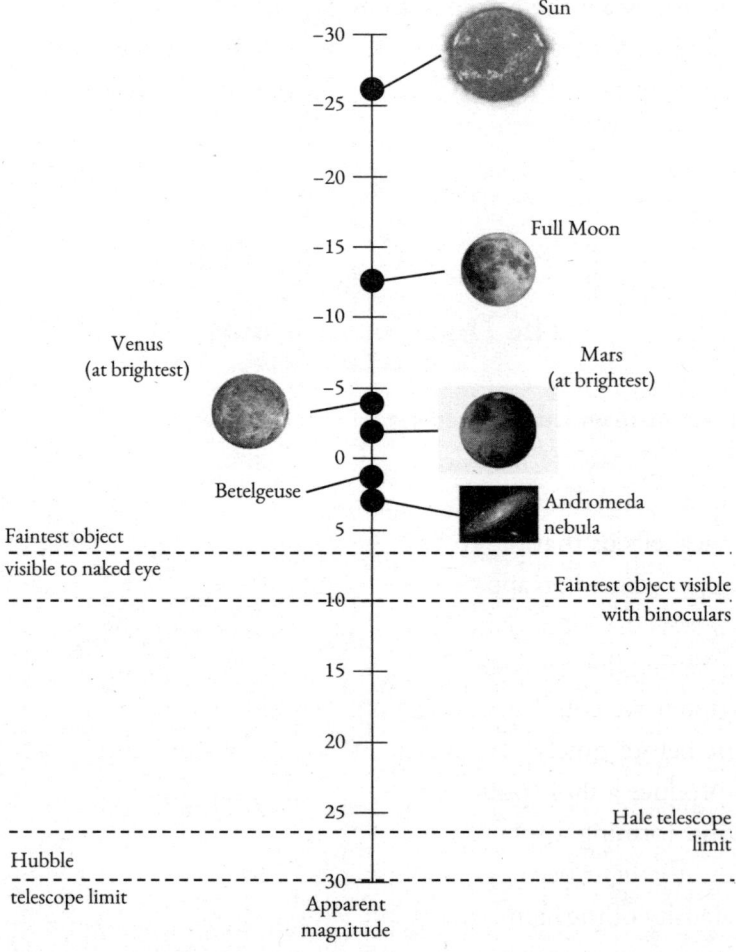

Fig. 2 The apparent magnitude scale runs backwards from the brightest objects (such as the Sun, $m = -27$) to the faintest objects visible to the Hubble telescope.

Alnilam, and Mintaka have apparent magnitudes of 1.75, 1.69, and 2.25, respectively. The red supergiant star Betelgeuse (Alpha Orionis), perched up on Orion's left shoulder, has an apparent magnitude of about 0.55 but this can vary from 0.2 to 1.2.

It takes some time to get used to a scale that works backwards. There are, of course, brighter objects in the universe than magnitude-1 stars, which means that the scale can reach through zero to negative numbers. On this scale the Sun has an apparent magnitude of −27 (see Fig. 2). At its brightest, the planet Venus has an apparent magnitude of −4.4. Astronomers denote such apparent magnitudes using the symbol m. This is just the first of many such symbols, and to help you keep track of these I've collected the most frequently used symbols in Appendix 1.

The Distance Modulus

The apparent magnitudes of stars tell us nothing about their intrinsic brightness or absolute magnitude. This is because the stars are positioned at different distances from our Earth-bound telescopes, and simple physics tells us that the further away a source of light, the dimmer it will appear. Based on apparent magnitude, Alnilam is the brightest of the three stars in Orion's Belt. Is this because it is intrinsically brighter or could it be that the star is simply nearer to us?

Imagine if we could turn a star off. We turn it back on for a brief moment before quickly turning it off again. We anticipate that this would produce a short pulse of light spreading out from the star in all directions, forming a sphere which expands outwards. In any specific direction, the light becomes diluted as the sphere grows bigger, as the fixed intensity of the light burst is spread over a greater and greater area. Now the surface area of a sphere is given by $4\pi d^2$, where d is the distance from the centre of the sphere to its surface. The brightness of a star a distance d from Earth is then diminished by a factor proportional to d^2.

This is an inverse-square law: a star three times more distant than some reference star with the same intrinsic brightness will appear nine times fainter.

There is nothing more we can do with this unless we have some means of measuring the distances. However, astronomers chose to define a scale of relative *absolute* magnitude, denoted M, and connected this with the apparent magnitude scale by defining M to be equal to m at a specific distance of 10 parsecs, or 32.6 light-years (for definitions of these distance measures, see Appendix 2). The difference $m - M$ is called the *distance modulus* and is often assigned the symbol μ (Greek mu). It is the difference between the apparent and absolute magnitude of a star. The relation is quite simple so I'll give it here:

$$\mu = m - M = -5 + 5 \times \log d.$$

In this expression d is measured in parsecs, and log means the logarithm in base 10. Note that if d is 10 parsecs, $\log d = 1$, $\mu = 0$, and M is equal to m. By introducing a third variable—the absolute magnitude—astronomers gave themselves two lines of attack. Knowing any two of these variables, say m and d or m and M, allows them to deduce the third.

For example, if we can find a way to measure the distance to a star (in parsecs), we can deduce its distance modulus. Measuring its apparent magnitude then allows us to determine its absolute magnitude. Alnitak lies about 820 light-years from Earth, or 250 parsecs. Its distance modulus is therefore about 6.99. An apparent magnitude of 1.75 (see above) implies an absolute magnitude of -5.24 (remember, the brighter the star, the lower—more negative—its magnitude). Similar calculations for Alnilam and Mintaka give absolute magnitudes of -6.38 and -4.99, respectively. Alnilam is intrinsically the brightest of the three.

Alternatively, and most importantly, if we can measure both the apparent and absolute magnitudes of a star we can determine its distance modulus and hence its distance (in parsecs), as 10 raised to the power

$(\mu + 5)/5$. The distance modulus for Alnilam is 8.07, from which we can calculate its distance as 411 parsecs, or 1340 light-years. A similar calculation for Mintaka gives a distance of 280 parsecs or 914 light-years. Alnilam is not only intrinsically the brightest star in Orion's Belt, it is also furthest away.

If some way can be found to determine the absolute magnitudes of observable stars (without actually having to go there), and if this measure can be *calibrated* for distance, then astronomers would have a yard-stick for determining the distances to the stars, and so measuring the universe.

Stellar Parallax

There is only one sure-fire way to measure astronomical distances and hence, from the equation for the distance modulus, the absolute magnitudes of stars. This involves the use of triangulation and relies on the phenomenon of *parallax*. A distant object will appear to shift position when viewed along two different lines of sight against a fixed background. Our human binocular vision exploits this parallax to create the perception of depth. In astronomy, the range of distance that can be measured depends on the length of the baseline of the triangle that can be drawn to a distant point, and hence the sharpness of the angle at its apex. The 'parallax angle' is half this.

Hipparchus made use of the Moon's parallax as measured from two different points on Earth's surface, and a little trigonometry, to determine that the distance from the Earth to the Moon lies somewhere between 36 and 40 times the Earth's diameter. Today we know that the diameter of the Earth is about 12,742 km at the equator, suggesting that the Moon is between 459,000 to 510,000 km distant. This is not a bad estimate, considering it was made over two thousand years ago. The Moon is actually about 384,000 km from the Earth, or about 30 times Earth's diameter.

This kind of parallax can't be used to measure distances even to nearby stars, as the baseline simply isn't long enough. But if we're patient, and have access to a telescope, then it becomes possible to measure stellar parallax by making observations at six-month intervals, when the Earth is at opposite sides of its orbit around the Sun. By this means, the baseline is extended to the diameter of Earth's orbit, which is about 300 million km. From these different perspectives, nearby stars are then seen to shift slightly against a background of more distant 'fixed' stars. This is an apparent motion due to stellar parallax, and a little trigonometry allows the distance to the star to be calculated from its parallax angle (see Fig. 3).

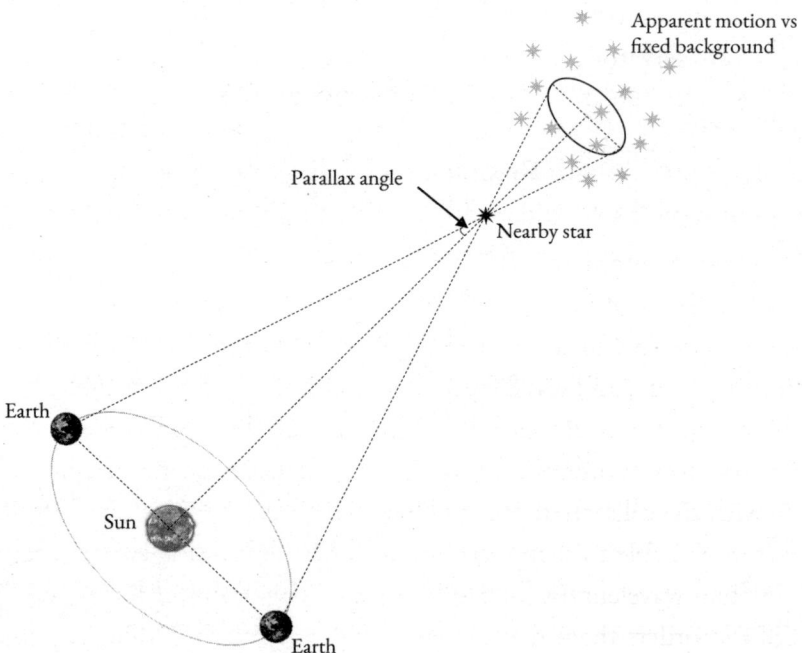

Fig. 3 The distances to nearby stars can be calculated by observing the apparent motions of these stars against the background of distant 'fixed' stars at opposite sides of Earth's orbit around the Sun. This is stellar parallax.

Astronomers define the parsec as the distance corresponding to a parallax angle of 1 arcsecond, or 1/60 of an arcminute, or 1/3600 of a degree, subtended by the distance between the Sun and Earth as shown in Fig. 3 (see Appendix 2).* If the parallax angle is small, then the distance in parsecs can be estimated simply as the reciprocal of the angle measured in arcseconds. The Sun's nearest stellar neighbour, Proxima Centauri, has a parallax angle of 0.769 arcseconds. To put this into some perspective, this angle is equivalent to that subtended by an object two centimetres in diameter located 5.3 km away. This parallax angle puts Proxima Centauri at a distance of 1.30 parsecs, or 4.24 light-years.

But the challenge posed by such measurements shouldn't be underestimated. By the end of the nineteenth century, parallax measurements of the distances to fewer than 100 nearby stars had been recorded, with different observatories often reporting widely different results.

Fraunhofer Lines

There is much more to starlight than its brightness. We know that when we disperse sunlight using a prism, the light is resolved into the familiar rainbow spectrum of colours. This procedure was well known in the seventeenth century and was famously described by Isaac Newton in a letter published in the *Philosophical Transactions of the Royal Society* on 19 February 1672: 'I procured me a triangular glass prism', he wrote, 'to try therewith the celebrated phaenomena of colours'.[6] In his experiments, Newton was able to demonstrate that white light is a mixture of light of different wavelengths (or colours), and the prism spreads these out in space and orders them from longer wavelengths (red) to shorter wavelengths (violet). What we perceive as red light has a range of wavelengths

---

* Fans of the *Star Wars* movies will know that Han Solo claimed to have completed the Kessel Run in the Millennium Falcon in less than 12 parsecs. As the parsec is a measure of distance, not time, some arm-waving was required in subsequent movies to help make sense of this.

from about 780 to 622 nanometres (billionths of a metre, nm). Violet light has wavelengths from 450 to 380 nm. Orange, yellow, green, blue, and indigo have ranges in between. Newton also observed that if the rainbow is allowed to pass through a further prism inverted relative to the first, the colours are mixed together again and white light is recovered.

Newton could not have observed that the rainbow spectrum is actually crossed by a series of dark lines, first discovered in 1802 by English chemist William Hyde Wollaston and independently rediscovered in 1814 by Bavarian physicist (and superb optical instrument-maker) Joseph von Fraunhofer. He found over 570, called 'Fraunhofer lines', using an instrument he developed called a *spectroscope* which allows the incoming light to be dispersed and the resulting spectrum to be observed through a viewing tube and eyepiece. He obtained similar results for the spectra of bright stars, but noted that the lines had different arrangements.

The Fraunhofer lines remained a bit of mystery until 1859, when the great German physicist Gustav Kirchhoff and German chemist Robert Bunsen (of burner fame) traced their origin to chemical elements such as calcium, hydrogen, iron, magnesium, and sodium in the outer layers of the Sun and stars (Fig. 4). Atoms of these elements do not absorb or emit light continuously across the red-to-violet spectrum of visible light, but instead absorb or emit light of very specific, narrow wavelengths— 'lines'—marching in patterns characteristic of each element. The dark Fraunhofer lines are the result of specific wavelengths of light that are absorbed by atoms of these elements, blocking them out, and so removing them from the continuous rainbow spectrum we observe on Earth. The explanation for such atomic 'line spectra' and their patterns would only come with the development of quantum mechanics in the mid-to-late 1920s, but astronomers didn't really need to know their origin. It was enough for them to understand that the lines could provide measures of the chemical compositions of the atmospheres of the planets, Sun, and stars.

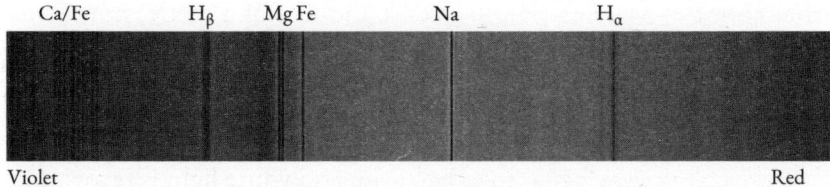

Fig. 4 Fraunhofer lines form a characteristic pattern across the spectrum of light emitted by the Sun. Each dark line can be traced to absorption of light by atoms of chemical elements such as calcium (Ca), hydrogen (H), iron (Fe), magnesium (Mg), and sodium (Na) in the Sun's outer layers. Here H_α and H_β refer to two lines in the absorption spectrum of hydrogen atoms.

Redshift

There was one further decryption technique available to early twentieth-century astronomers. A stationary police siren emits sound waves with a frequency of about 1800 Hertz, or cycles per second. On any average day it's more likely that you will hear a siren when a police car is racing to the scene of a road accident or chasing a perpetrator. As you stand on the side of the road, you will hear the siren's pitch change as the police car speeds past, a phenomenon known as the Doppler effect, named after the Austrian physicist Christian Doppler. This happens because the sound wave emitted in the direction of travel towards you becomes compressed—more cycles are squeezed into the distance between you and the siren. The incoming sound wave therefore has a higher frequency (higher pitch), and a shorter wavelength. The sound wave emitted in the opposite direction as the car speeds away from you becomes increasingly stretched out—the outgoing sound wave has fewer cycles per unit distance (lower frequency or pitch) and a longer wavelength (see Fig. 5).

Light, considered as a wave disturbance travelling through space, behaves in much the same way. One of the Fraunhofer lines can be identified as a specific line in the atomic spectrum of sodium, as Fig. 4 shows. This is the yellow sodium 'D' line (which is actually two closely spaced

blueshift

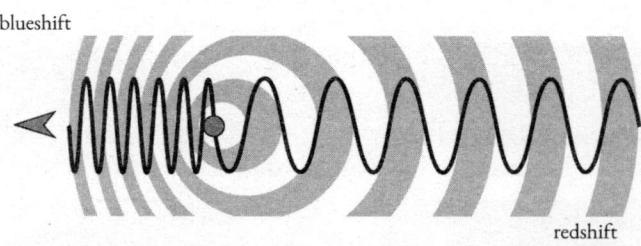

redshift

Fig. 5 Suppose a stationary object emits a sound or light wave with a certain frequency or wavelength. If the object is set in motion, the wave emitted in the direction of travel becomes compressed—there are more cycles per unit distance or the wave has a shorter wavelength (called a blueshift). The wave emitted in the opposite direction becomes stretched out—there are fewer cycles per unit distance or the wave has a longer wavelength (called a redshift). This is the Doppler effect.

lines) familiar from high-pressure sodium lamps used for street lighting. It has a wavelength of about 589 nm, as measured in a laboratory on Earth. Now suppose an astronomer is able to identify this line in the spectrum of a distant star. Because of the Doppler effect, we might expect that the line will be *shifted* relative to its position as measured on Earth. If the distant star is moving towards us, the wavelength will be shortened, just as the pitch of the police siren increases. This is called a *blueshift*—the line is moved to shorter wavelengths, towards the blue end of the spectrum. If the star is moving away, the wavelength of the line will be lengthened, called a *redshift*—the line is moved to longer wavelengths, towards the red (see Fig. 6).

The extent of the shift in the wavelength—towards the blue or red—in Fraunhofer lines observed in the spectra of distant stars can therefore tell us how fast these stars are moving through space towards or away from us.

The line spectra provide clues about the nature of the stars and their source of energy, their evolution in time, and the speeds at which they are moving. But stars were not the only objects of interest in the night sky. Astronomers of the late nineteenth century were also interested in the faint wisps of light that were too fuzzy to be classed as stars, called *nebulae*.

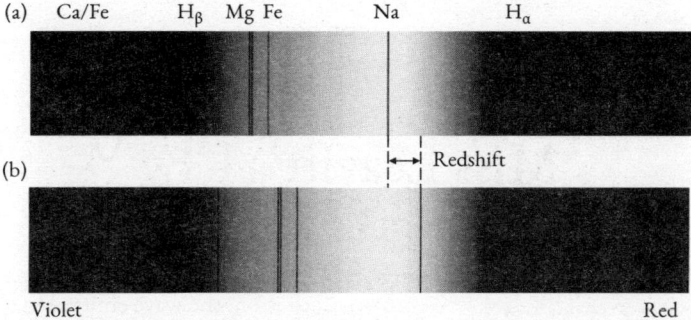

Fig. 6 The spectrum of light from the distant galaxy supercluster BAS11, located to the south of the handle of the Big Dipper, shows a redshift in its Fraunhofer lines (b) compared to the same lines in the Sun's spectrum (a). From the extent of the redshift, it is possible to deduce the speed at which the galaxy supercluster is moving away from us.

The Orion Nebula, located in the sword just to the south of the Belt, has an apparent magnitude of 4.0, and on a clear winter's night can be seen with the naked eye. The first photograph of the Orion Nebula (actually the first photograph of *any* nebula) was taken in 1880 by American doctor and amateur astronomer Henry Draper, who pioneered the use of photography in astronomy.

And this is where our story begins.

I

Miss Leavitt's Law

The Harvard College Observatory was founded in 1839, and in 1847 the Great Refractor, a refracting telescope with a 15-inch lens, then the largest of its kind in America, was installed in a separate domed building on Observatory Hill, a short walk from the Harvard University campus in Cambridge, Massachusetts. Though long retired, it's still there today.

In February 1877 Edward Pickering took up his appointment as the Observatory's fourth Director. His appointment caused a bit of a stir in the astronomy community, as Pickering was a physicist and there was no lack of experienced astronomers to choose from. But Pickering was a fastidious measurer. He had determined that the measurement of apparent magnitudes and the classification of stars according to their colour and spectra needed a firmer scientific basis, and chose to set as an objective for the Observatory the recording and cataloguing of the apparent magnitudes, colours, and spectra of all the observable stars. He was supported by generous donations from Mary Anna Draper, 'rich, red-haired, and a renowned hostess', the widow of Henry Draper, who had died in 1882. Mary had enthusiastically assisted in her husband's work.[1]

Theories about the origin, composition, and evolution of the stars could wait. Pickering judged that what astronomy needed most was lots and lots of accurate *data*. With Mary Draper's support, he established the *Henry Draper Memorial*, which would initially take the form of two extensive catalogues of stars classified according to their spectra and apparent magnitudes.

Pickering developed an instrument that made use of a clever combination of mirrors and prisms to bring any star accessible to a telescope into view alongside the North Star (or Pole Star, Polaris, or Alpha Ursae Minoris), chosen as a standard. Such a comparative technique is called *photometry*, and Pickering's instrument is known as a *meridian photometer*. Its use allowed the very rapid determination of the apparent magnitude of the star under investigation relative to that of the North Star. But, whilst this is perfectly fine for the northern sky, Pickering's ambition naturally extended to the southern sky too, and the North Star isn't visible in the southern sky.

Undaunted, in 1890 Pickering established a satellite observatory at Arequipa, about 8000 feet above sea level in southern Peru. It was named for Uriah A. Boyden, another sponsor who had bequeathed a substantial sum to support astronomy at Harvard. The station contained several instruments, including the 24-inch Bruce telescope, named for another benefactor, Catherine Wolf Bruce. Photographic plates were shipped around the tip of South America back to Boston, and thence to the Harvard observatory in Cambridge for analysis. By 1898 the British journal *Nature* was reporting that '... the equipment and maintenance of the observatory at Arequipa will be remembered as one of [Pickering's] most successful achievements'.[2] The Boyden Observatory was relocated in 1927 to Maselspoort, near Bloemfontein in South Africa, and the facility was donated to the University of the Free State in 1976.

The Harvard Computers

But it is, perhaps, a different achievement for which Pickering is best known. Whilst he had substantially improved efficiency in the gathering of data, he now faced a challenge in its subsequent analysis. This was something that couldn't be readily automated. The determination of the relative apparent magnitudes of stars—recorded as dark spots in the photographic negatives—still had to be done by eye. This involved mounting each plate on an angled wooden lectern set by a window so that sunlight would be reflected up through the plate by a mirror affixed to its base. Using a small magnifying glass or eyepiece, the analyser would then flit from dark spot to dark spot, assessing the brightness relative to a standard. It was tedious work, and not the kind of thing judged to be appropriate for qualified astronomers.

'A great observatory should be as carefully organized and administered as a railroad', declared Pickering in an address to the Harvard Chapter of the Phi Beta Kappa society in June 1906. He continued: 'A great savings [sic] may be effectuated by employing unskilled and therefore inexpensive labor, of course under careful supervision'.[3] This was Pickering's solution, although by 'unskilled' he really meant skilled, college-educated women who would work for 25 cents per hour, the minimum wage typical for unskilled workers. Among the first to be recruited was his housemaid, Williamina Fleming, who in 1899 would become Harvard's Curator of Astronomical Photographs.

Although Fleming was not college-educated, she had trained as a teacher in her native Scotland, and had become a maid after her husband had abandoned her and their young son in Boston. It was Pickering's wife Elizabeth who identified in her capabilities that reached well beyond those normally required of a housemaid. The 'Harvard Computers' would boast names now recognized for their extraordinary contributions to early twentieth-century astronomy, including Annie Jump Cannon, Antonia Maury, and Henrietta Swan Leavitt.

Spectral Types

In addition to the determination of apparent magnitudes, the Harvard Computers were also involved in classifying the different spectra of the stars. To this end Fleming had in the 1880s extended an earlier spectral classification system developed by Italian astronomer (and Catholic priest) Angelo Secchi. Fleming replaced Secchi's classes I through V with a letter classification A through P, with Q used to classify the spectra of odd stars that did not fit elsewhere. The basis for this scheme was the *intensity* of spectral lines identified as belonging to hydrogen atoms. Type A stars produce the strongest hydrogen lines whilst type O stars produce virtually no lines at all. Pickering did not concern himself with the possibility that the different spectral types might reflect physical differences between the stars, and so tell us something about their age and evolution. His principal concern was to gather and classify data and make these available to other astronomers.

In Fleming's scheme, Secchi's class I (white and blue stars including a subclass of white 'Orion' stars) became divided into types A, B, C, and D. Secchi's class II (yellow stars with weaker hydrogen lines but more prominent lines due to heavier elements such as calcium) was divided into types E through L. The Sun is a G-type star. Class III (orange-red stars such as Betelgeuse) became type M; class IV (red stars with strong bands and lines due to carbon) became type N; and class V (unusual stars with strong, broad emission lines) became type O. Pickering added a further type P, containing 'planetary' nebulae with bright line spectra. These are nebulae that happened to appear spherical to early astronomers (hence 'planetary').

Maury, a niece of Henry Draper who had studied chemistry and mathematics at Vassar, did not agree with Fleming's system. She believed that the detail now revealed in the spectra demanded a more nuanced system of classification. She devised a scheme consisting of 22 groups, each further subdivided into three categories based on the widths of the spectral lines: (a) average; (b) hazy; and (c) sharp. She was convinced that the sharp lines characteristic of her division (c) were an important distinction, writing '... it seems that stars of this division must differ more decidedly in constitution'.[4]

In 1868, Fraunhofer lines in the Sun's spectrum revealed the existence of a new chemical element, which was named helium (from *helios*, the

Greek word for Sun). The spectrum of helium was recorded in the laboratory for the first time in 1895, and this was subsequently identified as the source of the lines in the 'Orion stars'. Astronomers have a particularly idiosyncratic habit of classifying the elemental compositions of stars in terms of hydrogen, helium, and 'metals'. The last category includes all the elements heavier than hydrogen and helium. A star with high 'metallicity' is one with a relatively high proportion of heavier elements in its outer layers. The astronomers' 'metal' is of course very different from what we would normally consider to be a metal, and includes many non-metals.

The characteristic that would eventually unlock the mystery of stellar evolution was not the strengths of the spectral lines due to hydrogen, but the *surface temperatures* of the stars. There is an intimate connection between the temperature of an object and its colour. Heat any object to a sufficiently high temperature and it will emit light. We say that the object is 'red hot' or 'white hot'. Increasing the temperature increases the intensity of the light emitted and shifts it to a higher range of frequencies (shorter wavelengths). As it gets hotter, the object glows first red, then orange-yellow, then bright yellow, then brilliant white. To study such phenomena, theoreticians had adopted a simplified model based on the notion of a 'black body', a completely non-reflecting object that absorbs and emits light radiation perfectly. The intensity of radiation a black body emits is then directly related to the amount of energy it contains, and therefore its temperature. Wrestling with the theory of black-body radiation had in 1900 led German physicist Max Planck (unwittingly) to prepare the ground for the quantum revolution.

Maury's scheme ordered the stars according to their temperatures. Although her scheme was judged too complex to be useful, with Pickering's approval Cannon subsequently adjusted Fleming's classification system to reflect elements of Maury's design. Cannon reversed the positions of spectral types A and B, and dropped types now judged to be unnecessary. The result was a simpler system which classified spectra according to the colours and surface temperatures of the stars: B, A,

F, G, K, and M. Further decimal subtypes were introduced in 1912, for example type G now became G0 through G9 (the Sun is a type G2 star).

This left a small minority of stars in types O, R, and N. In 1913 the famed American astronomer Henry Norris Russell, director of the Princeton Observatory, argued that 'Of these O undoubtedly precedes B at the head of the series, while R and N, which grade into one another, come probably at its other end'.[5] The new system was immediately adopted and, with some small modifications, is still in use today. The brightest star in the triple Alnitak system is type O9.5, as is the primary star in the Mintaka system. Alnilam is type B0.

Cannon did not retain Maury's divisions (a), (b), and (c), but these were not forgotten. In 1905 the Danish astronomer Ejnar Hertzsprung had noticed that there appeared to be a relation between the absolute magnitudes of nearby stars, whose distances could be determined by observing their 'proper' motions in the night sky, and their colours. He identified a sequence from bright/white to dim/red. But there appeared to be two kinds of red stars: very faint stars of high density that fit the sequence and very bright stars of low density that did not. He was able to trace the difference to their spectra. The bright 'red giants' all exhibited sharp spectral lines—Maury's division (c). Noting that these divisions had by now disappeared from the spectral classification system, Hertzsprung protested to Pickering: 'To neglect the c-properties in classifying stellar spectra, I think, is nearly the same as if the zoologist, who has detected the deciding differences between a whale and a fish, would continue classifying them together'.[6]

Variable Stars

Leavitt joined the Observatory in 1893. She had the year before graduated from the 'Harvard Annex', the Society for the Collegiate Instruction of Women, which became Radcliffe College in 1894. If she had been permitted to attend Harvard University, her certificate would have been equivalent to a Batchelor of Arts degree, but women were not admitted to study at Harvard University until 1920. She had studied astronomy at the Observatory during her fourth and final year at the Annex, and must have found the subject sufficiently fascinating to want to continue after graduation, as an 'advanced student' engaged in what was effectively postgraduate research although, of course, there was no question of her qualifying for a PhD. The first woman to gain a PhD from Radcliffe College and Harvard College Observatory was British astronomer and astrophysicist Cecilia Payne, in 1925. Instead, Leavitt worked for free as a volunteer research assistant, no doubt much to the delight of the cost-conscious Pickering.

Leavitt was perfectly suited to the work. She had 'inherited in a somewhat chastened form the stern virtues of her puritan ancestors'.[7] She was a serious person, with a strong sense of duty, justice, and loyalty, with little time for frivolous distractions. She never married. Pickering put her to work on the analysis of the stack of photographic plates that was now rapidly accumulating. He was keen for her to establish new reference stars in both northern and southern skies and use these to determine the apparent magnitudes of all the others. This she did 'with unusual originality, skill, and patience'. She devised a simple method for making brightness comparisons. Small pieces of photographic plate holding images of reference stars were fitted with wire handles so that they could be moved around and held alongside the image under study. She called these 'fly spankers', as they looked like fly swatters but were too small to do a fly much damage.

She was also tasked with cataloguing variable stars, whose brightness changes periodically. At the time nobody understood why they do this, but then nobody understood anything about the physics of stars of any kind, and all astronomers could do was collect and catalogue the data. She began by looking at variable stars in the Orion Nebula. By studying photographs of individual variable stars taken at different times, it was possible to determine their periodicity, the time interval between peaks of maximum brightness or troughs of minimum brightness. This task was made easier by superimposing positive and negative photographic plates. The white spots of non-variable stars in the positive plate would be cancelled by the corresponding black spots in the negative. Variable stars would instead reveal a halo of whiteness or darkness which would grow larger in successive plates and then diminish as the star's brightness varied in time.

Leavitt worked diligently, and wrote up a draft paper on her results before embarking on a trip to Europe in 1896. She was gone for two years. On her return to America she went not to Cambridge but to Beloit, Wisconsin, to deal with some 'unexpected cares', family matters that she never properly explained. She also confessed to having problems with her hearing. Further waylaid by the illness of a family member, she did not return to the Harvard observatory until August 1902. She had remained in contact, exchanging letters with Pickering. But she didn't stay in Cambridge very long. She returned to Europe in January 1903, and was back at the observatory only much later that year, now as a permanent member of the staff with a salary of 30 cents per hour.

Under Pickering's direction, Leavitt turned her attention to variable stars in the Magellanic Clouds. As there is no equivalent of the North Star in the southern sky, the Portuguese explorer Ferdinand Magellan had instead relied for navigation on two bright, irregularly-shaped nebulae, one much larger (the Large Magellanic Cloud) than the other (the Small Magellanic Cloud—Fig. 7). Viewed through a telescope, these nebulae are observed to consist of multitudes of stars, including many variables.

As photographic plates continued to arrive from Arequipa, Leavitt identified more and more variable stars. By 1908, she had catalogued 1777 of them. She published her results in the *Annals of the Astronomical Observatory of Harvard College*. In this report she commented: 'It is worthy of notice ... that the brighter variables have the longer periods'.[8]

Fig. 7 The Small Magellanic Cloud as recorded by the ESA's Gaia satellite mission. This is not a photograph, but a composite produced by mapping the radiation detected by Gaia in each pixel.

A Remarkable Relation

This was the key. If the variable stars are all in the Magellanic Clouds then—in the big scheme of things—they are all roughly the same distance from the Earth, just as the inhabitants of Sydney are all roughly the same distance from Tower Bridge in London. Although Leavitt was measuring the apparent magnitudes (m) of the variable stars, the variation in magnitude can't be due to their distance and must therefore be due to variations in their *absolute* magnitude.

Look back at the expression for the distance modulus. If the distance d (and hence the distance modulus μ) is assumed to be the same for all the variables, then the differences in m must be due entirely to corresponding differences in M, since $m = M - 5 + 5 \times \log d$, and the distance d is fixed. In other words, the periodicity must arise because of the intrinsic physics of the variable stars, whatever this is. If periodicity could be connected directly with the absolute magnitude of a variable star—and if its distance could be *calibrated*—then any period–brightness relation (or period–luminosity relation, as it is called) would provide a means to measure the distances to any object in the vicinity of a variable star.

But in 1908 there were only 16 variable stars on which any relation between period and brightness or luminosity could be based. Leavitt judged these insufficient. She needed more data. Alas, in December 1908 she fell ill. She moved back to Beloit and although she was able to contribute to other research projects from her sick-bed, her recovery was agonizingly slow. She returned to Cambridge in May 1910, only to depart once more the following March when she learned of the death of her father. When she returned in the autumn of 1911, she was finally able to turn her attention back to the variable stars.

By March 1912, she had identified 25 variable stars in the Small Magellanic Cloud for which she was able to determine each individual star's maximum and minimum brightness and their period (called the 'light-curve'). She observed that for each star the brightness built quickly to a peak, which would then decline more slowly, before repeating. Their periods ranged from 1.25 days to 127 days. In a short paper communicated in the *Harvard College Observatory Circular*, Leavitt plotted the maximum and minimum brightness of each variable star against its period. The result was two curves—see Fig. 8(a). We've seen this kind of thing before, of course. The relation is *logarithmic*, and plotting maximum and minimum brightness against the logarithm of the period produces linear relationships, as shown in Fig. 8(b).

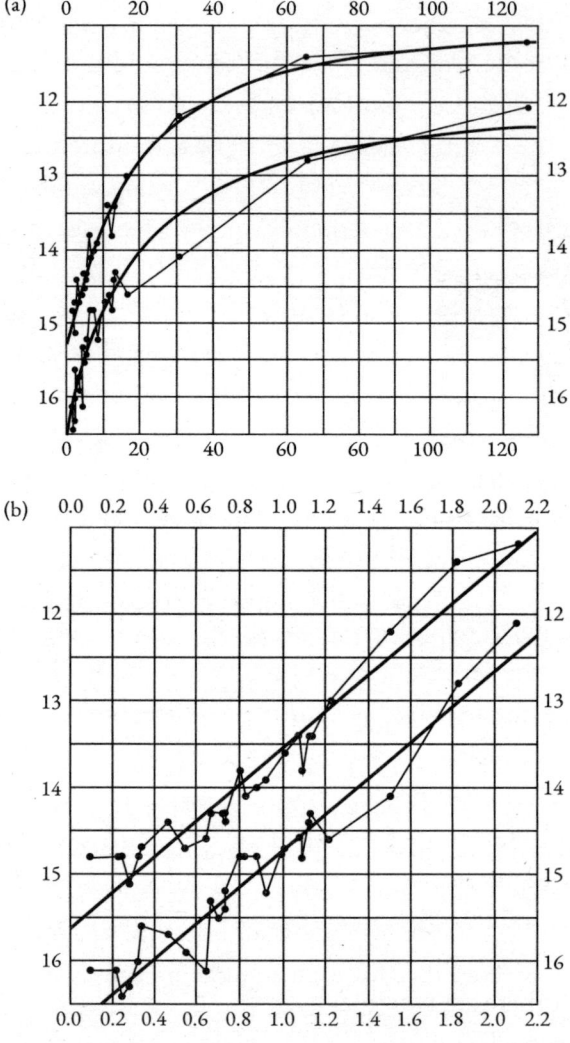

Fig. 8 These graphs are adapted Leavitt's original 1912 paper communicated in the *Harvard College Observatory Circular*. The graph at the top (Leavitt's Fig. 1) shows two curves corresponding to the maximum apparent magnitude (upper curve) and the minimum apparent magnitude (lower curve) plotted against the period in days. The graph below (Leavitt's Fig. 2) plots these magnitudes against the logarithm of the period. Note that in both graphs the magnitudes are plotted in reverse order, from high-to-low values (or from low-to-high brightness).

Photo 1   Henrietta Leavitt noticed a remarkable relation between the brightness of variable stars in the Small Magellanic Cloud and the length of their periods.

Leavitt wrote: 'A remarkable relation between the brightness of these variables and the length of their periods will be noticed'.[9] She continued: 'Since the variables are probably at nearly the same distance from the Earth, their periods are associated with their actual emission of light, as determined by their mass, density, and surface brightness'. But the period–luminosity relation that Leavitt had discovered, when combined with the distance modulus equation given earlier, can tell us only about *relative* distances. Unless the relation can somehow be calibrated.

Recall that there are two lines of attack. If the distance d to a variable star can somehow be measured, its absolute magnitude can be calculated from the apparent magnitude and the distance modulus. In this way, the relation between period and apparent magnitude can be turned into a relation between period and absolute magnitude. From this it is then possible to use measurements of the period and apparent magnitude of a variable star to deduce its distance. Leavitt and Pickering were well aware of this. Leavitt wrote: 'It is to be hoped, also, that the parallaxes of some variables of this type may be measured'.

Stellar Evolution

The years 1911–13 were to prove extraordinarily fruitful in the history of astronomy. In addition to Leavitt's period–luminosity law, the astronomical community would be gifted another device with which to decipher the messages hidden in starlight: the Hertzsprung–Russell (H–R) diagram.

In 1911 Hertzsprung had published a graph of absolute magnitude vs colour for stars in the Pleiades and Hyades. These are open star clusters containing stars of similar age, with very few giant stars outlying what would soon become known as the 'main sequence'. In 1913 Russell independently published an equivalent but more extensive graph of absolute magnitude vs spectral type. The English astronomer Arthur Eddington at Cambridge University obtained a version of this graph from Russell which he published a year later in his first book *Stellar Movements and the Structure of the Universe*. This is reproduced in Fig. 9, and would subsequently become known as a Hertzsprung–Russell diagram. These early diagrams plotted the spectral type from left-to-right (B, A, F, G, K, M, N), corresponding to *decreasing* surface temperature. The hotter, brighter stars feature top left, and the cooler, dimmer stars bottom right.

Although there is some considerable scatter, there could be no doubting the existence of a sequence. There also could be no mistaking the outliers, and Eddington likened the pattern to **Γ**, a mirror image of the number 7, with the main sequence running diagonally top left to bottom right, and a 'horizontal branch' running along the top.[10]

Russell speculated that stars began their lives as red giants, on the horizontal branch, contracting under their own gravity and dropping down

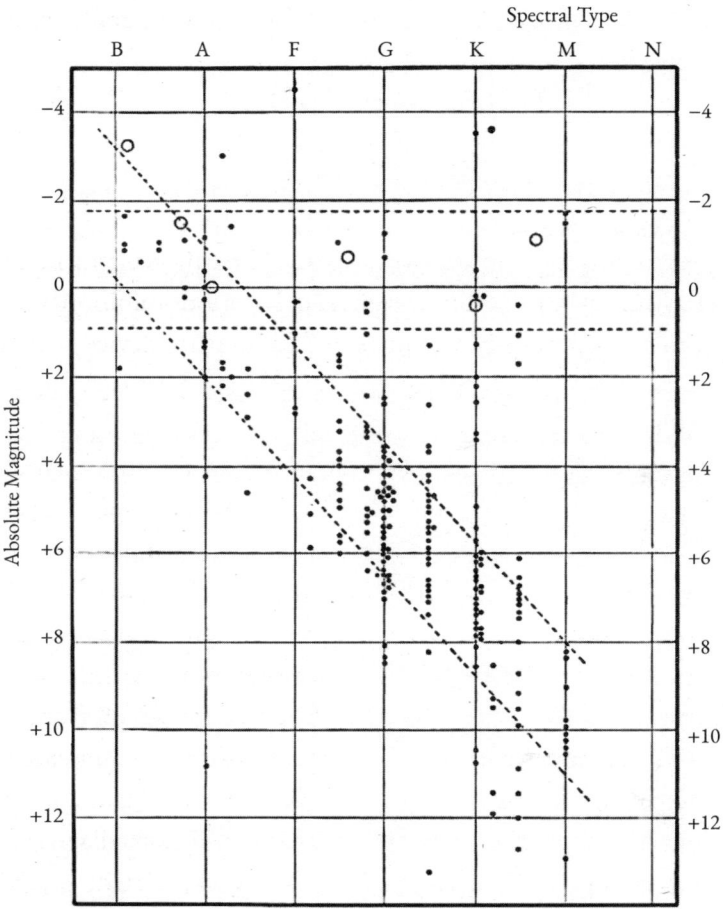

Fig. 9 Hertzsprung (in 1911) and Russell (in 1913) independently produced plots of the absolute magnitude of nearby stars vs colour or spectral type. This is a version of Russell's diagram that was reproduced in Eddington's *Stellar Movements and the Structure of the Universe* in 1914.

to join the main sequence. As this takes many millions of years it is not something that astronomers can watch as it unfolds. Rather, the diagram represents a snapshot of the stars as they appear to us now, and the main sequence represents the distribution of stars at different stages of their evolution. As they age the stars would be expected to lose their heat and become dimmer, descending the sequence towards the bottom right of the diagram.

Russell was almost right. The red giants are not, in fact, young stars on their way to the main sequence but rather old stars nearing the end of their lives that have branched upwards and away from the main sequence. Stars evolve in the opposite direction, from bottom right to top left of the diagram (and we can place the Sun—a type G star—roughly in the middle of the main sequence). In the denser main sequence stars, strong electric fields generated in the stars' outer layers cause the spectral lines to split. If such splitting is unresolved, the lines appear broadened or hazy. The outer layers of the much less dense red giant or supergiant stars have much weaker electric fields, and the spectral lines are consequently sharp: Maury's division (c).

Hertzsprung's Calibration

Everyone who studies maths at school learns that the equation of a straight line has the form $y = mx + c$. In this generalization we plot the values of y along the 'ordinate', or the vertical axis of a graph, and the values of x along the 'abscissa', or horizontal axis. The gradient or *slope* of the line is then given by m and the constant c is the point at which the line crosses the y-axis—c is the value of y when x is equal to zero.

Astronomers cast the period–luminosity relation discovered by Leavitt in the general straight-line form $\langle M \rangle = a + b \times \log P$. Here $\langle M \rangle$ signifies the average value of the relative absolute magnitude of a variable star and replaces the y-axis. The logarithm of the period (P, measured in days) replaces the x-axis, so b is the slope of the line and a is the constant, which astronomers call the *zero point*. The slope of the plot of brightness against the logarithm of the period is immediately available from Fig. 8(b). In her 1912 paper, Leavitt had noted that: 'The logarithm of the period increases by about 0.48 for each increase of one magnitude in brightness'. When rearranged this implies that the slope of a plot of increasing magnitude vs log P has the value -2.1.[*]

If the period–luminosity relation was to become a valuable method for measuring distance, then a way had to be found to calibrate it. One of the first tasks was to recognize that there is a difference between the human eye and an emulsion of silver halides (such as silver bromide or silver iodide), used for capturing images on a photographic plate. The latter are more sensitive to blue light (and less to red light) than the former, such that blue stars appear brighter—they have lower apparent magnitudes—in a photograph compared with naked eye observations. It was therefore necessary to distinguish two scales for apparent magnitude: photographic magnitude m_p, and visual magnitude, m_v.

Although this seems like yet another complication, it turns out to be quite a useful one, as the difference between m_p and m_v provided an early measure of a star's colour, and is called the *colour index*. Today the colour index of a star is defined as the difference in magnitude when observed through standard filters, designated UBV for ultraviolet, blue, and 'visual'. The 'visual' filter has a maximum transmission around 540 nm (green), corresponding to the peak in human visual perception.

The second, much more difficult, task was to find an accurate value for the zero point.

Measurements of stellar parallax were simply not possible for the variable stars that Leavitt had studied. But there is another way to make some estimates. The Sun is not motionless. When measured relative to what astronomers call the local standard of rest, the Sun moves through space with a 'peculiar velocity' of 13.4 km/s in the direction of the constellation Hercules. It carries the solar system along for the ride. Judged from the local standard of rest, the Earth's orbit around the Sun is displaced in space, describing a spiral.[†] The Earth doesn't return to the exact same point in space on completing its annual circuit.

---

[*] In effect, Leavitt's statement is equivalent to saying that $\log P + 0.48 \times \langle M \rangle = c$. We can quickly rearrange this to give $\langle M \rangle = a - 2.1 \times \log P$, where $a = c/0.48$ and we see that $b = -2.1$.

[†] If the Earth's orbit around the Sun were circular, the result would be a circular helix.

For stars observed within the Milky Way this effectively doubles the baseline measured at opposite sides of this elongated orbit. The analysis is complicated by the fact that the stars under observation are also moving through space relative to the local standard, and it's obviously important to distinguish between the apparent motion of the star (its displacement due to parallax) and its 'proper' motion. This uncertainty can be reduced if we determine the mean parallax and distance for sufficiently large collections of stars, since the velocities of the stars within the collection can be assumed to be distributed randomly and can be treated using statistics.

These challenges fell to Hertzsprung in 1913. He identified a collection of 13 variable stars within the Milky Way for which proper motions were known, including a prototypical variable called Delta Cephei in the constellation Cepheus, first discovered by the English amateur astronomer John Goodricke in 1784. Hertzsprung therefore chose to categorize these as 'Cepheid variables'. The name stuck.

He performed a statistical parallax analysis on the 13 Cepheids and concluded that a Cepheid variable with a period of 6.6 days should be about 7 absolute magnitudes *brighter* than the Sun. The distance to the Sun is 150 million km or 4.7 millionths of a parsec (Appendix 2), so its distance modulus is −32. We know that its apparent magnitude is −27, so its absolute magnitude is therefore about 5 (despite its importance to us, when seen from a distance of 10 parsecs the Sun is not a very bright star). So, according to Hertzsprung's analysis, a Cepheid variable with a period of 6.6 days would be expected to have an absolute magnitude of about −2.

He further applied a blanket colour index correction of 1.5 to Leavitt's photographic magnitudes. This helped him to set the zero point of the period–luminosity relation. Taking the slope of Leavitt's original plot allowed him to obtain the result:

$$\langle M \rangle = -0.6 - 2.1 \times \log P$$

In this equation, we take $\langle M \rangle$ to mean the average of the absolute visual magnitude.

This is quite extraordinary. We can apply this relation immediately—
as Hertzsprung did—to estimate the distance to the Small Magellanic
Cloud. Once corrected to apparent visual magnitudes, we can use Leavitt's data and Hertzsprung's calibration to calculate the absolute magnitudes and hence the distance moduli for each of the 25 Cepheid variables
that Leavitt had reported in 1912. There's inevitably some scatter but the
average distance modulus μ is of the order of 15. The distance is then
10 raised to the power $(\mu + 5)/5 = 4$, corresponding to 10,000 parsecs, or roughly 30,000 light-years. In his published paper, Hertzsprung
reported a parallax of 0.0001 arcseconds, which indeed corresponds to
10,000 parsecs.[11] And yet he converted this to just 3000 light-years. It's
not clear if this was simply an error on his part or a typographical error
that went uncorrected.

What is clear is that this was a distance measurement that was simply
unprecedented in the history of astronomy. It presaged a looming debate
about the size and nature of the universe. Were the Magellanic Clouds
and other nebulous objects that had been observed in the night sky part
of the Milky Way? Are the Milky Way and the universe one and the same
thing? Or do these objects lie outside?

Alas, in arriving at his calibration Hertzsprung had had to make some
big assumptions. A collection of 13 Cepheids is hardly sufficient to
count as a statistical sample, and he had applied the colour index correction uniformly to all the Cepheid variables reported in Leavitt's paper.
An important principle had been established, and a big first step had
been taken. But one key thread in the story that follows involves the
relentless and continuous improvement in both the sophistication of
measurement and the refinement of the period–luminosity relation. You
can get some sense for how far astronomers of the time had to go by looking up the distance to the Small Magellanic Cloud as it is measured today.
Hertzsprung was out by a factor of six. This distance is now known to
be about 60,000 parsecs, or 200,000 light-years.

There was much more work to be done.

2

The Scale of the Universe

Gazing up at a clear night sky, it's difficult to resist an overwhelming sense of stillness, and permanence. The stars shine, much as they have done for millennia, in the same immovable places in the sky, 'mansions built by Nature's hand'.[1] We may have read in articles and books about catastrophic events that can overtake stars in the last years of their lives, brightening to spectacular supernovae. But such events tend to be the preserve of astronomers, and very few of us ordinary mortals are ever likely to witness them. We could be forgiven for concluding that the stars and those faint, diffuse wisps we call nebulae sit motionless, their brightness suspended in the dark for all eternity.

But this could not be further from the truth. As astronomers in the early part of the twentieth century began to understand, all is in motion. Despite appearances, the universe is possessed of a restless energy. The key to this discovery was to be found once again in the examination of the light from distant objects in the sky, but now subjected to a very different kind of subtle instrumentation and analysis.

The Speeds of Spiral Nebulae

Percival Lowell founded the Lowell Observatory in Flagstaff, Arizona, in 1894, and became its first director. It remains very much a Lowell family enterprise, with the position of sole trustee of the Observatory passed down through the family.

A successful businessman, mathematician, and astronomer, for about fifteen years Lowell was obsessed with studies of the surface of Mars, detailing 'non-natural features' which he ascribed to canals and oases, and which he argued were the visible signs of an advanced Martian civilization desperate to source life-sustaining water from the planet's polar ice caps. Though strong on romance, Lowell's ideas were sadly weak on evidence. Some of the features he had considered as 'non-natural' were later identified to be perfectly natural geological features. Others are now thought to have been optical illusions. There is, however, plenty of evidence to suggest that the surface of Mars was once partly covered by water, long since lost to space.[2]

In 1901, Lowell recruited a young astronomer called Vesto Slipher, who had recently completed a Batchelor's degree in mechanics and astronomy at Indiana University. Though the position was temporary, Slipher was impressive and in 1903 went on to complete a Master's degree and, in 1909, a PhD at Indiana, on the subject of the spectrum of Lowell's favourite planet. At the time, Lowell had not wanted another permanent assistant and accepted Slipher only because he had already promised to do so, though he couldn't remember the length of term he had agreed. It hardly mattered. Slipher remained at the Lowell Observatory for 53 years, becoming its acting director on Lowell's death in 1916. He was formally named director ten years later.

The timing of Slipher's arrival at the Observatory's Mars Hill facility was auspicious. A new 24-inch refracting telescope, designed and built by Alvan Clark & Sons and costing $20,000, replaced an 18-inch telescope that had been loaned by another builder of precision instruments, John Brashear at the Allegheny Observatory in Pittsburgh, Pennsylvania. On packing up the telescope ready for its return, its 18-inch lens had been damaged. Brashear refused the offer of $400 in reparation, leaving Lowell in his debt. By way of compensation, it appears that Lowell agreed to purchase from Brashear the best spectrograph that money could buy.[3] Such a spectrograph is similar in principle to Fraunhofer's spectroscope, but with a camera in place of a simple viewing tube (the term spectrograph was first used by Henry Draper in 1876). This was delivered to Mars Hill shortly after Slipher's appointment. His first task was to mount it on the 24-inch telescope and figure out how to use it.

Slipher overcame some early problems with the instrument, and mastered it sufficiently to record high-quality spectra of Mars, Jupiter, and Saturn. Although his own research interests concerned the use of spectroscopy and the Doppler effect to measure the speeds of stars, his boss Lowell was more interested in the chemical compositions of planetary atmospheres and the rotation periods of the planets. Slipher accepted Lowell's priorities, pursuing his own interests only when circumstances permitted.

But in early 1909 Lowell directed him to study the spectra of a class of spiral nebulae that, at the time, were believed to be young solar systems in the making. Slipher was not optimistic, as the nebulae were very faint, but he nevertheless set to work. By now a master of the spectrograph, he realized that his ability to record a clear spectrum depended principally on the speed of the spectrograph's camera lens. By November 1910 he had rebuilt the instrument 'from equipment on hand', replacing the three-prism arrangement of Brashear's original with a single prism.[4]

In early December he recorded a faint spectrum of the archetypal spiral nebula, the Great Nebula in the constellation of Andromeda. He continued to refine the instrument and improve the quality of the spectrum through 1911 and 1912, just as Leavitt was painstakingly assembling her data on variable stars, two and a half thousand miles away at the Harvard College Observatory. In a series of long-exposure measurements performed in September, November, and December 1912 Slipher finally succeeded in recording spectra sufficiently detailed to determine the speed of the nebula from the Doppler shift of a group of Fraunhofer lines.

The result was not what he expected.

By this time the speeds of about 1200 stars and bright planetary nebulae had been measured and found to be of the order of tens of km/s, similar to the Sun's peculiar velocity of 13.4 km/s measured against the local standard of rest. This was the first time these techniques had been applied to determine the speed of a spiral nebula, but there was no good reason to think that the result would be much different. Instead, Slipher found that the spectrum of the Andromeda Nebula was substantially blueshifted and moving with an unprecedented speed of 300 km/s.

This result caused Slipher some considerable doubt, and for a time he even questioned the applicability of the Doppler shift as an indicator of speed. But he was soon reconciled. He wrote: 'The magnitude of this velocity, which is the greatest hitherto observed, raises the question whether the velocity-like displacement might not be due to some other cause. I believe we have at present no other interpretation for it. Hence we may conclude that the Andromeda Nebula is approaching the solar system with a velocity of about 300 km/s'.[5]

Lowell thought that Slipher had made a great discovery. 'Try some other spiral nebulae for confirmation', he suggested.[6]

Slipher did just this. By April 1913 he had found that Fraunhofer lines in the spectrum of a spiral nebula in the constellation Virgo showed a substantial redshift, with an utterly astonishing speed of about 1000 km/s. In August 1914, at the 17th American Astronomical Society meeting in Evanston, Illinois, he reported results for 15 nebulae with speeds up to 1100 km/s, observing that: 'As well as may be inferred, the average velocity of the spirals is about 25 times the average stellar velocity'.[7] Three years later he had collected data on 25 spiral nebulae. Only four were blueshifted, with speeds ranging from 30 to 300 km/s. The remaining 21 were all redshifted, with speeds ranging from 150 to 1100 km/s.

In March 1914, Hertzsprung had written to Slipher suggesting that his results held enormous implications for our understanding of the scale of the universe. By his own estimate from the year before, the Small Magellanic Cloud must lie outside the Milky Way. Could it be that the spiral nebulae are actually unresolved stellar systems, or 'island universes', also lying far outside the Milky Way? Hertzsprung argued that their speeds strongly suggested the affirmative. Slipher, more comfortable with the measurements themselves and wary of their interpretation, may have judged that any such conclusion was premature.[8] But in the paper announcing his results published in the *Proceedings of the American Philosophical Society* in 1917 he wrote:[9]

> It has for a long time been suggested that the spiral nebulae are stellar systems seen at great distances. This is the so-called 'island universe' theory, which regards our stellar system and the Milky Way as a great spiral nebula which we see from within. This theory, it seems to me, gains favor in the present observations.

Einstein's Universe

Just a few months before Slipher delivered his paper to the American Philosophical Society, Albert Einstein wrote from Berlin to his friend and colleague, Austrian theorist Paul Ehrenfest. 'I have ... again perpetrated something about gravitation theory which somewhat exposes me to the danger of being confined in a madhouse'.[10]

Einstein was by now an established figure with an international reputation. In 1905 he had published what would become known as his special theory of relativity (and the iconic equation connecting mass and energy $E = mc^2$, though in his original paper he writes this as $m = E/c^2$). And, although few in the physics community would acknowledge this just yet, he had also laid the foundations for the new theory of quantum mechanics. Ten years later, his general theory of relativity finally eliminated Newton's mysterious (if not occult) force of gravity, which had seemed to act instantaneously and at a distance. In what Einstein called his field equations of gravitation, this force is replaced by a reciprocal action between matter (or mass–energy) and four-dimensional spacetime.

In Minkowski spacetime (named for German mathematician Hermann Minkowski), the three dimensions of space are combined with time multiplied by the speed of light (which has the same dimensions as a distance) to form a four-dimensional *spacetime metric*. A large quantity of matter (such as a star, or planet) is understood to curve the space-time around it. Nearby objects are caught out by this curvature, and we mistake their resulting free-fall as a force of attraction. In Einstein's theory (and with acknowledgments to American physicist John Wheeler), matter tells spacetime how to curve, and curved spacetime tells matter how to move. In very simplistic terms, Einstein's field equations can be summarized like this:

$$\text{Spacetime curvature} = 8\pi\,G \times \text{Mass-energy}$$

In this expression G is Newton's gravitational constant.

But any theory of matter and energy, and space and time, is also potentially a theory of the entire universe. Just two years after triumphantly presenting his general theory to the Prussian Academy of Sciences, Einstein was back with a new theory of *physical cosmology* which, he judged, might risk confinement in a madhouse.

Einstein first had to decide the kind of universe on which to apply his equations. He conceived of a universe which is spatially finite (not infinite), but which has no edge—a universe which is finite but 'unbounded'. At first sight this seems logically impossible, but only if we imagine the universe to be flat, like a plate. We know well enough that the Earth is finite—it has a mass that can be estimated (it's about six million billion billion kilograms). And, although from any human vantage point on its surface the Earth looks flat, we know that it is near-spherical, and so has no edge. From our perspective on its surface, the Earth is finite but unbounded.

So, in Einstein's universe spacetime curves back on itself much like the surface of a sphere, and contains a finite amount of matter. The 'sphere' in question is a sphere in four spacetime dimensions (what mathematicians refer to as a 3-sphere or, more generally, a hypersphere). The surface of the Earth, in contrast, is a 2-sphere, a sphere in three-dimensional space. But a similar logic applies. If it were possible for us to set off together across Einstein's universe on a journey along a straight path, we would eventually find ourselves returning to our point of departure. Eddington applauded Einstein's move: 'I think Einstein showed his greatness in the simple and drastic way in which he disposed of difficulties at infinity. He abolished infinity. He slightly altered his equation to make space at great distances bend round until it closed up'.[11]

Einstein had further to assume that the universe is uniform in all directions, containing objects that have the same kind of composition. He also had to assume that the universe we observe from our vantage point on Earth is no different from the universe as observed from any and all such vantage points. In other words, observers on Earth occupy no special or privileged position. What we see is a 'fair sample' of the universe as a whole. This is the *cosmological principle*.

He now had to deal with a more stubborn problem. Gravity (understood in terms of the curvature of spacetime) draws objects together but it doesn't push them apart. Put another way, matter curves spacetime only in ways which cause objects to move towards each other, never away.

Newton had understood where this leads. The mutual gravitational attraction between all the objects in an infinite Newtonian universe would eventually cause the universe to collapse in on itself. He had had no choice but to invoke the ultimate solution, arguing in the *Mathematical Principles of Natural Philosophy* that God had placed the stars sufficiently far apart to prevent such a catastrophic collapse from happening.[12] But, just as in Newton's gravity, there was nothing in Einstein's equations that would help to hold the universe steady. Einstein needed a physical mechanism that would do much the same job that Newton had asked of God.

This could not be fixed with another assumption, and he chose instead to fudge the equations. The left-hand side of Einstein's gravitational field equation describes the extent of spacetime curvature that will determine the motions of all the matter in the universe, which is summarized on the right-hand side. To engineer a universe that stands still, he attempted to balance the equation by introducing a new term on the left-hand side, imbuing spacetime with a curious kind of anti-gravity, a negative gravitational pressure which builds in strength over large distances, counteracting the effects of the curvature caused by all the matter on the right-hand side:

Spacetime curvature − 'Cosmological term' = $8\pi\,G$ × Mass-energy

This new 'cosmological term' is characterized by a constant, now typically given the symbol Λ (Greek lambda). By carefully selecting the value of Λ, Einstein thought that he could achieve perfect balance: a static universe. In this 'Einstein universe', an object with a mass m will experience a repulsive force roughly equal to $m \times \Lambda \times d$, where d is its distance from some point of observation. A suitable choice of size for Λ means that the repulsive force becomes relevant only at very large distances. He admitted that he: '... had to introduce an extension to the field equations that is not justified by our actual knowledge of gravitation'.[13] But at the time this must have seemed like a neat solution.

General solutions of Einstein's field equations do not distinguish between space and time dimensions, but of course it is our experience that the universe consists of three spatial dimensions and one time dimension. If the universe is assumed to be homogeneous in both space and time, we are free to disentangle spacetime to examine the structure of space at a particular instant in time. Einstein's universe is spherical in four spacetime dimensions. But if we separate space and time and suppress two of the spatial dimensions, we can show how the remaining dimension—a line—evolves in time.

Because Einstein's universe is 'closed', the line is wrapped around into a circle, but we need to keep in mind that only the circumference of the circle represents this one-dimensional space, such that if we want to populate this space with a few objects, these must be placed on the circumference. Einstein's universe is also static, which means that the circumference of the circle does not change in time. If we plot snapshots of the circle taken at different times along the time axis, and join these

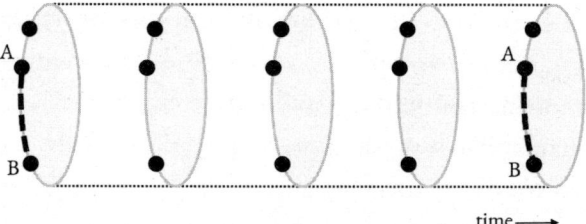

time⟶

Fig. 10 In Einstein's closed, static universe, or 'cylinder world', space is fixed and unchanging in time. In this figure 3-dimensional space is shown as the 1-dimensional circumference of a circle, and objects embedded in this space are therefore shown on the circumference. We can then observe how 'snapshots' of space change through intervals in time, marching left to right. Note how these objects (such as A and B) maintain their spatial relationships with each other through time.

up, the result looks like a cylinder, as shown in Fig. 10. Einstein's static universe is sometimes referred to as a 'cylinder world'.

This might all seem perfectly reasonable, but in formulating the gravitational field equations to achieve a static universe, Einstein had made some significant errors. Not only was there no evidence or even a real theoretical justification for the cosmological term, it actually doesn't do quite what Einstein claimed. The universe that he described in his 1917 paper is actually rather unstable, and quite capable of tripping over into expansion (in which the circumference of the circle increases with time) or contraction (in which it decreases with time).

Shapley's Calibration

In a paper presented at the 15th meeting of the Astronomical and Astrophysical Society of America, held over the New Year period from 31 December 1912 to 2 January 1913, Russell reported his own determination of the absolute magnitudes of some nearby Cepheid variables. He performed a statistical parallax analysis from a preliminary general catalogue of the proper motions of stars assembled by the American astronomer Lewis Boss. But he did not consider Leavitt's results or Hertzsprung's calibration.[14] This challenge would fall to one of Russell's Princeton graduate students.

Harlow Shapley had wanted to study journalism. He had worked as a crime reporter for newspapers in Kansas and Missouri, and had figured that a formal qualification might bring him more interesting assignments. He had enrolled at the University of Missouri in 1907, only to discover that the new School of Journalism wouldn't open its doors for another year. 'So there I was, all dressed up for a university education and nowhere to go', he wrote in his autobiography.[15]

Undaunted, he reached for a catalogue which listed courses alphabetically. He claimed that he couldn't pronounce 'Archaeology', so settled

Photo 2 Harlow Shapley had wanted to study journalism, but on arriving at the University of Missouri in 1907 he discovered that the new School of Journalism wouldn't open for another year. He consulted a course catalogue, claimed that he couldn't pronounce 'Archaeology', so settled for the next on the list, 'Astronomy'. Shapley (left), pictured conversing with Ejnar Hertzsprung at the Moscow Meeting of the International Astronomical Union.

for the next on the list, 'Astronomy'. He graduated in 1910 and, after completing a Master's degree a year later, he won a prestigious fellowship at Princeton University, where he worked on eclipsing binary star systems under Russell's supervision. The work caught the attention of the astronomy community and, as he was finishing his thesis, Shapley met with George Ellery Hale, 'the founder of almost everything astronomical'.[16] Hale was director of the Mount Wilson Observatory in the San Gabriel Mountains near Pasadena, California, home to the 100-inch Hooker reflecting telescope, then the world's largest. Shapley was nonplussed: they didn't discuss astronomy at all. It was enough for Hale to learn that Shapley was a 'decent guy', and shortly afterwards he was appointed to a position at Mount Wilson.

In 1914 Shapley published a paper on the origin of the periodicity of the Cepheid variable stars, arguing in favour of an intrinsic (though still obscure) pulsation rather than the popular alternative hypothesis involving eclipsing binary systems, in which the observed variation in brightness is caused by one star passing in front of the other. Four years later he presented his own extended period–luminosity calibration.

There are essentially three components to Shapley's calibration. He made use of the same data as Hertzsprung, but rejected two of the Cepheid variables that Hertzsprung had considered in 1913, which he judged to be 'not typical'.[17] The end result was not too different from Hertzsprung's original analysis. For a Cepheid variable with a mean period of 5.96 days Shapley estimated an absolute magnitude of -2.35 ± 0.19. Compare this with Hertzsprung's calibration, which gives -2.23.

To accommodate Leavitt's data on the apparent magnitudes of the much more distant variable stars in the Small Magellanic Cloud, Hertzsprung had converted her photographic magnitudes to visual magnitudes by applying a blanket colour index correction of 1.5. Shapley's correction was a little more sophisticated and varied with the period: $-0.55 + 1.5 \times \log P$. This gives corrections ranging from 0.2 for a

Cepheid variable with a period of 1.25 days to 2.0 for a Cepheid variable with a period of 127 days. Shapley was satisfied that the same period–luminosity relation applies for the nearby, short-period variable stars studied by Hertzsprung and Russell and the longer-period variables studied by Leavitt.

But even this was already in error. There are many reasons why a distant bright object appears faint. Distance may be the principal reason, but the object may also be veiled by interstellar gas and dust. A star seen through such a mist will appear dimmer than it really is and the further away it lies the greater the effect. Underestimating the apparent magnitude means that, for a given absolute magnitude, the distance modulus is overestimated, and hence the distance is overestimated.* The rotation of the object may also have an effect on its proper motion, as judged from Earth. And, of course, a limited data set leaves the analysis exposed to both random and systematic error.

Shapley now unwittingly compounded the error by including different kinds of variable stars observed in globular clusters, spectacular spherical assemblies of hundreds of thousands or millions of stars that form a spherical 'halo' around the Milky Way's central bulge. The largest of these globular clusters is Omega Centauri, visible to the naked eye, known in antiquity and identified as a cluster of stars by Edmond Halley in 1677. Solon Bailey at the Harvard College Observatory had catalogued the periods of many variable stars in Omega Centauri, and many more in other globular clusters, and had commended them to Shapley.

The variable stars in the globular clusters are distinctly dimmer and have shorter periods than the Cepheids, but they held the advantage of being much more numerous. Although outnumbered, Shapley was nevertheless able to identify Cepheids in Omega Centauri and two other nearby clusters, and he used these and his modified period–luminosity

---

* Remember, $\mu = m - M$, so if m is underestimated (measured to be larger or less negative than it really is) then μ is overestimated and from $d = (\mu + 5)/5$, d is overestimated.

relation to determine their distances. This gave him a fix on the luminosities of the globular cluster variables, which he then used to determine
the distances to another four globular clusters. In this way Shapley was
able to extend the period–luminosity relation to more than 230 stars
(see Fig. 11). This was no longer a simple straight line but a curve,
with the absolute magnitudes of the shortest-period variables appearing to become independent of period. Instead of fitting the result to
the equation of a straight line, Shapley provided a table which could
be used to read off absolute magnitude in increments of tenths of
$\log P$.

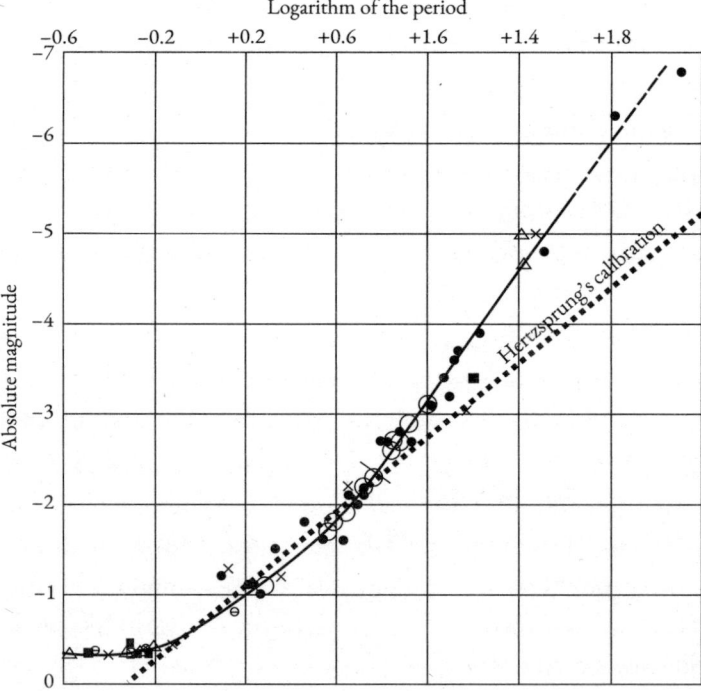

Fig. 11 Shapley's calibration of 1918, adapted from Fig. 1 of his paper in the
Astrophysical Journal. For comparison with Fig. 8 (Leavitt's Fig. 2) I've reversed
the horizontal axis which in Shapley's original ran from high to low values of the
logarithm of the period. I've also added a reference to Hertzsprung's 1913
calibration. The different symbols represent data from seven different star systems.

It would transpire that the error introduced by including the variable stars in the globular clusters almost exactly compensated the error in the original statistical parallax analysis. This meant that the globular cluster variables were judged to have absolute magnitudes close to zero, much as we would expect today, and as later studies confirmed these expectations, confidence in Shapley's calibration grew. The errors would not be discovered for another thirty years.

The Great Debate

Shapley now used his already rather rickety distance ladder to estimate the luminosities of 30 of the brightest stars in each cluster, rejecting the brightest five in case these were stars lying in the foreground, between Earth and the cluster. He found that the average luminosities of the remaining 25 stars were similar for each of the seven clusters, so he used these as standard candles to estimate the distances to another 21 clusters. From these distances he estimated the apparent diameters of the clusters, providing a kind of standard rod which he used to determine the distances to a further 41 clusters. The end result was an extraordinary 'map' of the distribution of globular clusters surrounding the Milky Way.

In 1901, the Dutch astronomer Jacobus Kapteyn had published a 'first approximation' of the scale of the Milky Way and, as common understanding at the time had it, the scale of the entire universe. By 1920 the 'Kapteyn universe' was an oblate collection of stars, slightly thicker and denser at its centre, about 60,000 light-years in diameter and 8000 light-years thick. Kapteyn judged that the Sun lay close to the centre.

But Shapley's map of the globular clusters appeared to tell a very different story. If, as Shapley surmised, the clusters are arranged spherically around the central bulge of the Milky Way, then their distances suggested that the Sun is far from the centre. Shapley estimated a distance of 50,000 light-years. He further estimated the dimensions of his universe to be a staggering 300,000 light-years in diameter with a central bulge 30,000

light-years thick. Shapley's universe engulfed the Magellanic Clouds. His
fellow Mount Wilson astronomer Adriaan van Maanen had measured
the rotation speeds of some spiral nebulae, and if these were indeed separate island universes as large as Shapley's Milky Way this implied that
they were rotating at speeds much faster than light.[18] In deducing his
special theory of relativity, Einstein had assumed that the speed of light is
a maximum that cannot be exceeded. The spiral nebulae must therefore
be smaller, and closer.

Shapley's radical model of the universe caught the attentions of
astronomers, and dissent was quickly fomented. Amongst the most vocal
was Heber Curtis, at the University of California's Lick Observatory on
Mount Hamilton, in the Diablo Mountain range near San Jose. Shapley's Milky Way was greatly over-sized. Curtis favoured a much smaller
galaxy, some 30,000 light-years in diameter and 7000 light-years thick at
its centre (and so smaller than Kapteyn's universe). The spiral nebulae
must lie firmly outside the Milky Way, at vast distances of 500,000 to 10
million light-years, island universes of the same order of size as the Milky
Way itself.

As the noise grew, Hale made a passing remark at a Council Meeting
of the National Academy of Sciences. Perhaps an evening during a scheduled Academy meeting in April 1920 could be set aside for a debate, paid
for from a fund set up to honour Hale's father, either on the subject of
the island universe theory or on Einstein's relativity. The Academy eventually settled on the island universe theory, to be debated by Shapley and
Curtis.

It would later be exalted as the 'Great Debate'. The protagonists gathered in the auditorium of the Smithsonian's US National Museum in
Washington, DC at 8:15 p.m. on 26 April 1920. Both gave talks entitled 'The Scale of the Universe'. These were necessarily short, and they
were poorly matched. Shapley's talk was very general and elementary,
and so for the most part uncontroversial. Pickering had died in February 1919, and Shapley was being considered as a potential successor to
the directorship of the Harvard College Observatory. Curtis, the more

experienced lecturer and public speaker of the two, gave a rather technical talk about spiral nebulae. By July he was installed as director of the Allegheny Observatory in Pittsburgh. Both published much more detailed papers back-to-back in the May 1921 *Bulletin of the National Research Council*.

Although Curtis is often promoted as the winner of the debate, in truth there was no clear winner. There is a sense in which both were right, and both were wrong. Curtis' Milky Way was too small, Shapley's too large. Shapley was right that the Sun lies out in the suburbs of the Milky Way, and Curtis was right that the spiral nebulae are island universes. But the debate itself was inconclusive. It had settled nothing.

The argument was resolved rather emphatically in 1925, by another Missourian: Shapley's fellow Mount Wilson astronomer Edwin Hubble.

Major Hubble

Hubble's journey to astronomy was in some respects more tortured even than Shapley's. Hubble's fascination with the planets and stars had begun at age eight, with his first encounter with a telescope built by his grandfather. They were both especially fascinated by the planet Mars, and Lowell's fanciful interpretations of features on its surface. But Hubble's absorption with science and the stars drew displeasure from his father, a strict Calvinist who judged that pursuit of a scientific understanding of the universe was an insult to its divine creator. Though Hubble never lost his interest in astronomy, at high school this became submerged by both his father's tyrannical rule and a curriculum biased towards the classics. Hubble also possessed an athletic prowess that put him in the school record books.

His efforts on and off the field earned him a scholarship to the University of Chicago in 1906 where, at his father's insistence, he took courses in law. But he also prepared for the possibility of a career in astronomy by studying science (majoring in physics) and mathematics. The professor of physics and department head was Albert Michelson, who in 1907 became the first American scientist to be awarded a Nobel prize, for his experimental studies of the speed of light. Michelson protégé and assistant professor Robert Millikan was just turning his attentions to experiments on the elementary charge of the electron that would earn him a Nobel prize in 1923. Hubble worked for a time as a student assistant in Millikan's laboratory.

In 1910 Hubble realized a long-standing ambition, and won a three-year Rhodes Scholarship to Oxford, England, with a handsome stipend of $1500 per year (nearly $50,000 today). He was free to choose his research discipline, but exercising this freedom would mean direct confrontation with his father's wishes. He elected to study law and international law at Queen's College.

Within just a few days of his arrival he had undergone a remarkable transformation, affecting English mannerisms and adopting a rather dubious accent. He took to wearing plus fours, a tweed jacket, and a flat cap, escaping Oxford at weekends for a 'ripping good time'.[19] When not laughing at his 'effort to acquire an extreme English pronunciation', one of his fellow Americans at Oxford recalled that he was 'full of astronomy ... and not terribly interested in jurisprudence'.[20] He spent a lot of time at Oxford's Radcliffe Observatory, where he would amaze the astronomers working there with a knowledge of contemporary astronomy quite unexpected of a law student (who detested the law). Those rather enamoured of Hubble included Herbert Hall Turner, the observatory's director, credited with coining the term 'parsec'.

Hubble completed his BA in 1912, leaving him with nine months of his scholarship to spend studying whatever he wished. Although he had spent considerable time in the company of astronomers, and was highly regarded by Turner, he first elected to study literature, before quickly switching to Spanish.

Hubble was still in England when his father died in 1913. This was an emotionally heavy blow but one that offered relief from his father's overbearing manner, and his insistence that astronomy was an unsuitable career for his son. But the weight of his father's influence pulled on his conscience from beyond the grave. With his father's passing his older brother and sister struggled to pay accumulated debts and keep the family whole. Hubble returned to America to pick up the pieces. He never qualified to practise law in the US (despite later claiming otherwise), working instead as a freelance translator of legal documents for companies doing business in South America. He was subsequently hired by New Albany High School in Indiana to teach Spanish, physics, and mathematics, and to coach the basketball team.

It was only in May 1914 that he decided to reach out to his former astronomy professor at Chicago, asking if it might be possible to secure financial support to study for a PhD. At his professor's urging, Hubble wrote also to Edwin B. Frost, director of Chicago's Yerkes Observatory. Hubble secured a modest tuition scholarship, and started at the Observatory the following August. At the 17th American Astronomical Society meeting in Evanston, he heard Slipher talk about his observations of the extraordinary speeds of spiral nebulae.

Hubble's path was now set.

But his dissertation, titled 'Photographic Investigations of Faint Nebulae', written and examined in a rush in 1917, was unimpressive. He had applied for a commission with the Officers Reserve Corps and in September 1918, sixteen months after enlisting, Major Hubble set off back across the Atlantic aboard an overcrowded army transport. He never saw combat (despite later claiming otherwise). The armistice with Germany was signed in a railroad carriage in Compiègne, France, at 11 a.m. on 11 November 1918. Hubble stayed on with the occupation forces in France and Germany, before moving on to Cambridge, England, where he had an opportunity to renew his acquaintance with astronomy. Hale had pledged in 1917 that, on his return to America, Hubble could take up a position at Mount Wilson. He sailed to New York in August 1919, and was discharged from the army in San Francisco later that month. He made a brief stop at the Lick Observatory on his way south, and joined the staff at Mount Wilson on 3 September.

The combative Shapley did not much care for Hubble. His fellow Missourian affected an aristocratic air and spoke 'Oxford', which he judged to be a cover for his limited achievements in astronomy. Shapley found him cold, and unapproachable: '[Hubble] just didn't like people', he claimed, '[H]e was a Rhodes scholar and he didn't live it down'.[21] When Shapley departed for the Harvard College Observatory in 1921, Hubble was glad to see him go.

The Realm of the Nebulae

Milton Humason had been eleven years old when in 1902 he first climbed the trail up Mount Wilson, riding on the back of a mule. Thus began a love affair with the mountain that would last a lifetime. Five years later, he joined the team of 'muleskinners', tasked with transporting construction materials, equipment, and supplies to the summit of the mountain, where the Mount Wilson Observatory was being built. Humason was a high school dropout, with no formal education or trade to fall back on. As the construction project wound to a close, he joined a citrus ranch in the foothills of the mountain with the intention of learning the kinds of ropes that would keep him in gainful employment, and allow him to marry. His father-in-law, keen to ensure his daughter's future, was the observatory's chief electrician and engineer. By 1915, Humason and his wife Helen had saved enough money to buy their own citrus ranch in Monrovia, a few miles east of Pasadena.

In October 1917, Humason's father-in-law had advised him that an opportunity to join the staff at the observatory had arisen. The janitor was leaving, and the position was available. The meagre salary was offset somewhat by free housing on the mountain. As an added incentive, the job included training as a night assistant on the large reflecting telescopes. Humason started the job in November, just as the 100-inch Hooker telescope saw 'first light'. Encouraged by Shapley and other astronomers, he busied himself with the task of learning as much as possible about advanced astronomy, becoming a uniquely talented observer, and making himself indispensable. By 1919 Hale 'thought highly' of his abilities.[22] A year later he was being put to work as an astronomer in all but name.

Just before Shapley's departure for Harvard in 1921, Humason had approached him with a couple of photographic plates of the Andromeda Nebula. Humason had thought the plates showed a Cepheid variable, and had marked its position on the plates. Could this be used to determine the distance to the Andromeda Nebula? Shapley stuck to the position he had adopted for the Great Debate, and assured him that the nebula lay firmly within the Milky Way. He proceeded to wipe the plates clean of Humason's marks.[23]

Humason's efforts were finally rewarded with the official title of astronomer in 1922 and, in October 1923, Hubble found the Cepheid variable that Humason had first discovered in Andromeda. Hubble had been using the 100-inch reflector to search for novae, characterized by a sudden burst of brightness followed by a decline over about 100 days or so. But he quickly realized that one suspected nova was pulsating, with a period of 31.415 days. Hubble used Shapley's calibration to convert the period into an absolute magnitude. Subtracting this from the photographic magnitude gave him the distance modulus, and hence a distance of at least 300,000 parsecs, or nearly a million light-years. Shapley's calibration was still full of errors, and we now know that this distance is nearer 2.5 million light-years. But, even in error, this distance was more than three times the diameter of Shapley's bloated universe. By February 1924, Hubble had found nine novae and two Cepheid variables in Andromeda.

He wrote to Shapley to tell him the 'good' news. Cecilia Payne, who would go on to become the first woman to gain her PhD from Harvard College Observatory, was in the room with Shapley when he opened the letter. He held it out to her, remarking 'Here is the letter that has destroyed my universe'.[24]

But Shapley had his reputation to consider and did not readily concede. Hubble just ploughed on, finding Cepheids in other spiral nebulae, placing them well beyond the boundaries of Shapley's Milky Way. Hubble was wary of rushing his results into print, not least because of the conflict with van Maanen's rotation speeds. His paper 'Cepheids in Spiral Nebulae' was read by Russell at a meeting of the American Association for the Advancement of Science held in Washington, DC on 1 January 1925. Hubble didn't attend, but Shapley was in the audience.

The mood of the astronomy community changed almost overnight in favour of the island universe theory, and Shapley had to admit defeat. Van Maanen's rotation speeds were in error. 'They wonder why Shapley made this blunder', he later explained. 'The point is that the reason he made the blunder is that van Maanen was his friend and he believed in friends!'[25]

The scale of the universe increased to accommodate objects at previously unheard-of distances, moving at extraordinary speeds. Hubble's was a remarkable achievement, but throughout his life he refused to call Andromeda and other spiral nebulae 'galaxies'. He would always refer to them as nebulae. He later wrote: 'The conquest of the Realm of the Nebulae is an achievement of great telescopes. It began with the identification of nebulae as independent stellar systems, comparable with our own system of the Milky Way'.[26]

But it was also an achievement of great astronomers.

3

Hubble's Constant

One of Einstein's principal motivations for developing his cosmological theory in 1917 was to realize an ambition set by Austrian physicist Ernst Mach. In Newton's 'classical' (pre-quantum) mechanics, space and time are taken to be absolutes, providing a fixed, static backdrop against which events play themselves out. Objects move in an absolute space, marching to an absolute time set by some mysterious cosmic clock. Take all the objects out of Newton's universe and we are obliged to suppose there would still be an empty 'container' of some kind.

Mach rejected this view, arguing that an absolute time without reference to physical change has '... neither a practical nor a scientific value; and no one is justified in saying that he knows aught about it. It is an idle metaphysical conception'. He continued: 'No one is competent to predicate things about absolute space and absolute motion; they are pure things of thought, pure mental constructs, that cannot be produced in experience'. Mach preferred to ground mechanics more firmly in our direct experience of physics, based on what we know and free of 'pure mental constructs': 'But if we take our stand on the basis of facts, we shall find we have knowledge only of *relative* spaces and motions'.[1]

In an appendix added to later editions of a popular book explaining his theories of relativity to the wider public, Einstein credits Mach as the only physicist '... who thought seriously of an elimination of the concept of space, in that he sought to replace it by the notion of the totality of the instantaneous distances between all material points. (He made this attempt in order to arrive at a satisfactory understanding of inertia.)'[2]

Left to its own devices, an object will continue in a state of rest or uniform motion in a straight line unless it is acted on by some force: Newton's first law of motion. Why? The presumption is that the object possesses inertia, a natural resistance to changes in its speed and/or direction of motion. Newton called it *vis insita*, an innate force of matter.[3] A short time spent on thrill-rides at a fairground convinces us of our own inertia, which we experience directly as we are suddenly accelerated, and spun around. Rotational motion involves a constant change in direction for the object rotating, and so is a very particular form of acceleration. Newton had argued that the centrifugal forces arising from rotational motion appear to be independent of the object's immediate surroundings, and must therefore be the result of absolute motion. And if absolute motion exists, so must absolute space. To deny this is to insist that rotational motion must rather be relative. But then we are obliged to ask: relative to what?

Mach was somewhat ambiguous in his response. But phrases such as '... such [centrifugal] forces *are* produced by [an object's] relative rotation with respect to the mass of the earth and the other celestial bodies',[4] were taken by Einstein to mean that inertial forces result from relative motion with respect to the mass of the entire universe. In 1918, Einstein elevated this presumption to the status of a principle, and argued that his cosmological theory (suitably modified by the cosmological term) fulfilled the requirements of 'Mach's principle'.[5] Inertia is relative. It results from all the matter (mass–energy) in the universe. Gravity results from local deviations from the homogeneous distribution of matter, which create pockets of spacetime curvature. A universe containing a single object would therefore be presumed to possess no inertia.

De Sitter's Universe

Dutch theorist Willem de Sitter, professor of astronomy at Leiden University, didn't agree. Einstein believed that any valid solution of his gravitational field equations would demand a universe containing at least some matter. But just a few months after publication of Einstein's paper in 1917, de Sitter showed that there is a valid solution, in which space is also positively curved, which contains no matter at all. It is literally an empty container. This solution was not meant to be taken seriously as a model of our own universe, which obviously does contain matter, but de Sitter was looking to make a point. Einstein's universe implies the existence of what de Sitter called 'world-matter', which fills the universe. 'It is, however, also possible to satisfy these equations without this hypothetical world-matter. Then, of course, the "material postulate of [the] relativity of inertia" is not satisfied, but the "mathematical postulate", which makes no mention of matter ... is satisfied'.[6]

De Sitter's universe had some curious properties. Introduce a (massless) observer and a 'test particle' into de Sitter's universe and it immediately becomes dynamic. The test particle doesn't stay put, as might be expected in a 'static' universe. Instead, the observer will be aware that the particle accelerates away into the emptiness. Its acceleration increases with distance, such that any light emitted from it will appear increasingly redshifted, until it disappears at the horizon. This is because time in de Sitter's universe runs more slowly with increasing distance from the observer, so that light from even a stationary test particle will appear redshifted, the extent of the shift increasing with increasing distance.

It works like this. The frequency of light (Greek nu, ν) is the number of complete up-and-down cycles per unit time. If time runs more slowly at larger distances, the unit measure becomes *longer* relative to the observer's reference. The number of cycles become stretched out, such that the observer sees a *lower* frequency. The wavelength (Greek lambda, λ) is related to the frequency according to: $\lambda = c/\nu$, where c is the speed of light. If the speed of light is constant, a lower frequency therefore means a larger (longer) wavelength, or a redshift.

This was all rather intriguing. Such a redshift actually has nothing to do with the Doppler effect—light from a distant test particle *at rest* would still be redshifted, to an extent which depended on its distance from the observer. It was known that Einstein's general theory of relativity predicts a time dilation effect in the vicinity of a large mass, such as the Earth. Light moving away from a massive object therefore becomes redshifted. But there were no large masses in de Sitter's universe so this could not be a gravitational redshift, either. This was something else entirely. It became known as the 'de Sitter effect'.

That de Sitter's empty universe was contrived and artificial did not necessarily mean that it was incapable of providing some physical insight, and de Sitter was aware of the possibilities. He had gained his doctorate in 1901 under the supervision of Kapteyn in Groningen in the Netherlands, which had included studies of the moons of Jupiter at Cape Town Observatory in South Africa. With Kapteyn he had contributed to research on the structure of the Milky Way galaxy, determining the parallaxes and apparent magnitudes of nearby stars. He had been appointed professor of astronomy at Leiden in 1908. He was well aware of the results of Slipher and others on the redshifts (and hence speeds) of distant spiral nebulae.

He wrote that, in his universe: 'The lines in the spectra of very distant stars or nebulae must therefore be systematically displaced towards the red, giving rise to a spurious positive radial velocity'. 'Spurious' because if the test particle is at rest, then quite obviously it's not moving. This is not a Doppler effect.

De Sitter's universe made use of a rather complex coordinate system, and it was not at all clear if the de Sitter effect was an artefact of this. But he noted that the very large velocities reported for the spiral nebulae are inconsistent with the observed speeds of stars in our immediate neighbourhood, and the majority appear to be moving away, precisely as demanded in his model universe. He speculated that 'If ... continued observation should confirm the fact that the spiral nebulae have systematically positive radial velocities, this would certainly be an indication to adopt the hypothesis B [de Sitter's universe] in preference to A [Einstein's universe]'.[7] In essence, he was suggesting that the large speeds of distant spiral nebulae might have a *cosmological* origin. This was the first attempt to connect models of the universe derived from Einstein's general theory of relativity to astronomical observations of the speeds and distances of spiral nebulae.

Einstein was horrified by de Sitter's solution and rejected his arguments. Yes, this was a valid solution of the gravitational field equations but it was purely mathematical, and of no relevance to the physics of our universe. As far as Einstein was concerned, without matter there can be no spacetime. Because de Sitter had used a rather obscure coordinate system, he wondered if, in fact, the matter necessary for the existence of spacetime was not all hidden at the horizon, equivalent to what would later become known as an 'event horizon' of the kind identified with black holes.

De Sitter's challenge nevertheless had the desired effect. Mathematicians and mathematically inclined physicists and astronomers were drawn to the study of the two solutions, in efforts to determine which was more satisfactory. The redshift data for the spiral nebulae were deemed too limited and uncertain to provide any kind of empirical test, and their origin remained mysterious. The attentions of astronomers were focused instead largely on the fallout from the Great Debate.

Friedmann's Dynamic Solutions

Prevailing opinion had led to the presumption that our universe is static and, at least on a large scale, unchanging. The interest in Einstein's and de Sitter's universes reflected this presumption. Indeed, in 1922 Russian physicist and mathematician Alexander Friedmann showed that these are the *only* static solutions. But he also showed that there are further valid solutions of the gravitational field equations that contain matter and are not static, but *dynamic*. They describe universes in which space expands and contracts, or which cycle endlessly between expansion and contraction.

Friedmann had studied mathematics and physics at St Petersburg University, graduating in 1909. His education had included seminars on quantum mechanics and relativity from Ehrenfest. A master's degree in pure and applied mathematics followed, which included studies in aeronautics and meteorology. During the first world war he served in the Russian air force and developed theories of precision bombing. He became head of the Central Aeronautical Station first in Kiev, then in Moscow. When this was disbanded following the October Revolution in 1917, he taught mechanics at Perm University before returning to St. Petersburg (by this time renamed as Petrograd), where he founded a school of theoretical physics.

In two papers published in 1922 and 1924, Friedmann explored different kinds of dynamic solutions to Einstein's gravitational field equations. These were exercises in mathematics. Like de Sitter, he did not seek to represent our own universe but, unlike de Sitter, he made no reference to astronomical data, other than to declare their inadequacy. Although he chose to retain the cosmological term that Einstein had introduced, he assumed that it could take any value, including zero.

He discovered that the properties and behaviours of the universes represented by his solutions depend on the relationship between the amount of matter, which tends to curve spacetime and so draw itself together, and the size of the cosmological constant Λ, and hence the size of the cosmological term, which tends to push the matter apart. In Einstein's static universe, space is necessarily positively curved (the one-dimensional line representing space in Fig. 10 curves and closes to form a circle) and unchanging in time. Such spatial curvature is assigned a parameter k. This is a 'dimensionless' parameter, meaning that it is a pure number without associated units (such as kg, or s). In Einstein's static universe $k = +1$.

In his paper of 1922, Friedmann showed that there are solutions with $k = +1$ in which space can expand or contract. A universe is said to be 'closed' when the density of matter is high, with many objects in a given volume of space, and the size of the cosmological term ensures a modest rate of expansion. In a closed universe, parallel lines will cross at some point and the angles of a triangle will add up to more than 180°, as shown in Fig. 12(a). A closed universe will expand for a while before slowing, the surface of the sphere in Fig. 12(a) getting larger and larger. But the high density of matter means this expansion will at some point grind to a halt, before space turns in on itself and starts to collapse. Friedmann found that a universe with a cosmological constant of zero will oscillate back and forth, alternating between expansion and contraction, the period of oscillation depending on the amount of matter in it.

In a semi-popular book, *The World as Space and Time*, published in Russian in 1923, Friedmann summarized the results of his mathematical explorations. Although the absence of reliable astronomical data

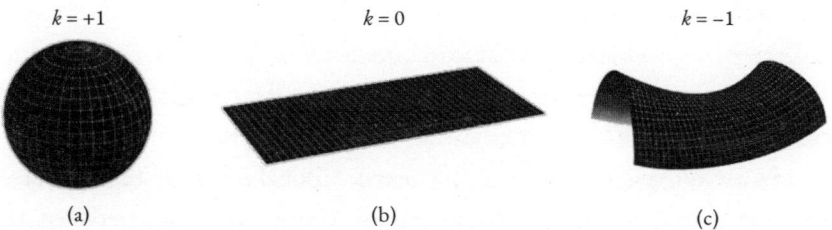

Fig. 12 Different solutions of Einstein's gravitational field equations possess differently curved spaces. In 1922 Friedmann showed that there are finite dynamic solutions in which space is 'non-Euclidean'. A 'closed' universe, which expands for a time before collapsing, and has $k = +1$, a positively curved space, (a). In 1924 he presented solutions corresponding to an infinite 'open' universe, which expands forever, and has $k = -1$, a saddle-shaped, negatively curved space, (c). In 1925 Howard Robertson presented solutions with $k = 0$, corresponding to flat, Euclidean space.

rendered 'useless' the task of associating numbers with his solutions, he couldn't resist speculating 'for the sake of curiosity' on the time that had elapsed since the universe 'was created starting from a point to its present state', for which he volunteered 'tens of billions of our ordinary years'.[8] Although he showed no preference for any particular solution, he seems to have been fascinated by the possibility of a cyclical universe.

The following year he published a further paper in which he examined solutions in which space is negatively curved, with $k = -1$. Such universes are infinite, they are 'open' and space will expand forever. In this kind of open universe, parallel lines will again cross and the angles of a triangle will add up to less than $180°$, Fig. 12(c).

In his 1925 PhD thesis, the American physicist Howard Robertson described a metric with zero spatial curvature, $k = 0$, corresponding to familiar 'flat', three-dimensional Euclidean space, Fig. 12(b). Space in such a universe returns us to a more familiar schoolroom geometry: parallel lines stay parallel and the angles of a triangle add up to $180°$. Robertson subsequently published his results in 1927. He would go on to devise a general metric that is both homogeneous (of the same kind everywhere) and isotropic (the same when measured from different directions), and which can take positively curved, flat, and negatively curved forms.

Ehrenfest had been appointed as professor of physics at Leiden University in 1912, and Friedmann had stayed in contact with him. In April 1922 he had sent Ehrenfest an earlier manuscript titled 'On the Geometry of Curved Spaces'. Einstein, by now a Nobel laureate, became aware of another of Friedmann's papers titled 'On the Curvature of Space' shortly after it was published that same year. He wrote a short commentary published in the same journal rejecting Friedmann's solutions as 'suspect' and incompatible with his field equations.[9]

Alerted to Einstein's objections by his colleague, Yuri Krutkov, Friedmann wrote to Einstein in December 1922 asking that he look again at his calculations and issue a correction. Einstein discussed Friedmann's letter with Krutkov and Ehrenfest in Leiden in May 1923, and was persuaded that he had indeed been mistaken. He submitted a retraction shortly afterwards. He now found Friedmann's solutions 'both correct and clarifying'. But it seems he was still far from convinced. In his hand-written submission to the journal he accepted that the field equations admit dynamic solutions '... to which a physical significance can hardly be ascribed'.[10] He thought better of it, and chose to strike out the end of this last sentence before sending it to the editor.

It also seems that Friedmann himself did not realize the extent to which his solutions might be relevant to the physical cosmology of our own universe. As far as he was concerned, this was all an exercise in math-ematics and the available astronomical data were too unreliable to be informative. It's likely that Friedmann's perceptions would have changed as more astronomical data were gathered and reported. But in August 1925, on his return from a holiday in the Crimea with his new wife Natalia, he fell ill after eating some unwashed pears. He had contracted typhus, but this went undiagnosed. He died on 16 September, aged 37.

A young Ukrainian student called George Gamow, who was intend-ing to study for a PhD with Friedmann, was obliged to find another thesis advisor. Krutkov came to his rescue.

Lemaître's Expanding Universe

Friedmann's papers had no impact on the astronomical community, which continued to puzzle over the speeds of spiral nebulae. Although Slipher had contributed more nebulae to his data set, he had not sought to publish them, preferring instead to make them available to other astronomers. Eddington was one such beneficiary. Eddington had famously led one of the solar eclipse expeditions in May 1919 to deter-mine the extent to which light from distant stars is bent as it passes by the Sun, thus providing observational evidence in support of general relativity. In his book *The Mathematical Theory of Relativity*, pub-lished in 1923, Eddington provides a table 'kindly prepared for me

Photo 3 A gathering of stellar physicists at Leiden Observatory. Back row, standing left to right: Albert Einstein, Paul Ehrenfest, and Willem de Sitter. Front row, sitting: Arthur Eddington and Hendrik Lorentz.

[by Prof. V.M. Slipher at the Lowell Observatory] ... containing many unpublished results'.[11] The table lists data for 41 nebulae, of which only 7 are blueshifted. The velocities for redshifted nebulae ranged from 150 to 1800 km/s.

In 1922 the German astronomer Carl Wirtz suggested that there might be a formal relationship between speed and distance, and a year later the German mathematician Hermann Weyl suggested that for small distances such a relationship might be linear, with speed directly proportional to distance (as in $y = mx$). But in 1924 the Swedish astronomer Knut Lundmark published a graph of speed against distance, and concluded that the data were too scattered to draw a definitive conclusion. Fellow Swede Gustaf Strömberg reached similar conclusions a year later. He had used the same data set of 41 nebulae supplied by Slipher, supplemented by data for a further two nebulae reported by others, the Magellanic Clouds, and 18 globular clusters.

The situation changed rather dramatically in 1926 when Hubble published a detailed statistical study of 400 nebulae. This was principally an exercise in classification, dealing with magnitudes, dimensions, and distances. Hubble found an inverse-square correlation between the diameters of the nebulae and their apparent magnitudes, suggesting that the nebulae are all of roughly the same absolute magnitude, so that their apparent magnitudes arise principally from their distances. By calibrating the relationship using nebulae whose distances were known (for example from studies of Cepheid variables) he was able to connect the distance d of a nebula to its apparent magnitude, m: $\log d = 0.2m + 4.04$. Note that apparent magnitude refers here not to individual stars but rather *entire nebulae*. Hubble's paper did not deal with redshifts and speeds. He had not yet turned his attention to the possibility that there might be a relationship between speed and distance.

The de Sitter effect was undoubtedly a clue to the origin of the enormous speeds of the spiral nebulae, but de Sitter's choice of coordinate system meant that the situation was rather confused. In 1925, Belgian physicist Georges Lemaître identified the problem. In Einstein's cylinder world (Fig. 10), space is positively curved and the universe is static. Space is also positively curved in de Sitter's universe, and a one-dimensional line can again be pictured as a circle. In Einstein's universe time flows uniformly (from left to right in Fig. 10) through all points in space. But de Sitter had chosen coordinates in which time flows at different rates through different points on the circumference of the circle. In this system it is possible to identify a central origin point, through which time flows at its fastest, but for points further and further away from this time flows more and more slowly.

This results in a static universe, but de Sitter's system offends the principles on which Einstein had built his original cosmological theory in 1917. A centre to the universe offers a unique perspective, and means that space is not uniform in all directions (Lemaître called it a 'spurious inhomogeneity'). Observers at different locations in de Sitter's universe will record different observations of the same phenomena.

Lemaître sought to fix this problem by changing the coordinate system, restoring a uniform flow of time though all points in space. There were two immediate consequences. Homogeneity was recovered, as required, but the resulting universe was now dynamic—space could expand or contract. Unlike Einstein, Lemaître saw some promise in an expanding universe: 'Our treatment evidences this non-statical character of de Sitter's world which gives a possible interpretation of the mean receding motion of spiral nebulae'. But the simple change of coordinate system had led him to a Euclidean space which, to Lemaître's mind, must be infinite in all directions. Infinite space implies an infinite amount of matter: 'We are led back ... to the impossibility of filling up an infinite space with matter which cannot but be finite'.[12]

An expert in both astronomy and Einstein's general theory of relativity, Lemaître was well placed to sort this out. He had begun to study for a degree in civil engineering at the Catholic University of Louvain in 1917, interrupting his education to serve as an artillery officer in the Belgian army during the first world war. He resumed his studies after the war, switching to mathematics and physics, and gaining a doctorate in mathematics in 1920. He was ordained as a Catholic priest in 1923. A thesis on general relativity earned him a scholarship to study abroad that same year, and he chose to study with Eddington in Cambridge.

Eddington introduced him to astronomy and cosmology. From Cambridge he moved to Harvard College Observatory to work with Shapley, and registered for a second doctorate (on relativity) with the Massachusetts Institute of Technology (MIT) in Boston. Whilst in America he became acquainted with Slipher and Hubble. He was subsequently appointed to a professorship at Louvain and returned to Belgium in 1927.

He also returned to the problem of de Sitter's universe. This was clearly not a good starting point and, as he was unaware of Friedmann's 1922 solutions of the gravitational field equations, he was obliged to rediscover them for himself. In 1927 he found that the equations admit solutions with non-Euclidean, positive spatial curvature, avoiding the need to fill infinite space with a finite amount of matter. The solutions are similar to Einstein's cylinder world, except that they are dynamic: the radius of the universe (or the radius of space reduced to the circumference of a circle) can increase or decrease.

Objects placed on the circumference now appear to move apart or move closer together, and the further away they are from each other the faster they *appear* to be moving. As the spiral nebulae nearly all appeared to be moving away as judged from our vantage point on Earth, the obvious conclusion is that the universe must be expanding, as depicted in Fig. 13. In Lemaître's expanding universe the cylinder of Einstein's static universe is replaced by a shape called a *hyperboloid*.

In de Sitter's universe, light from a test particle is redshifted both because of its motion away from the observer (a Doppler effect) and because time slows down with increasing distance from the observer (a cosmological effect). This works a little differently in Lemaître's expanding universe. Suppose a distant spiral nebula is actually stationary in relation to an Earth-bound observer. In such a case there can be no redshift associated with the Doppler effect. But it will take many years for light from a distant nebula to reach us, and in the time it takes for this light to travel to the Earth the space in which it is travelling will have expanded. We can see this immediately from Fig. 13. If light sets off from

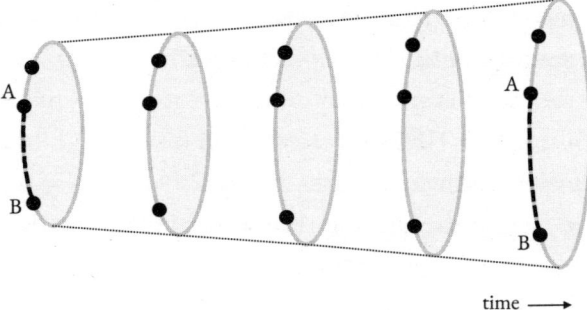

time ⟶

Fig. 13 Like Einstein's static universe (Fig. 10), Lemaître's expanding universe is closed (as before, 1-dimensional space is represented by the circumference of a circle). But unlike Einstein's universe, it is expanding—its radius increases with time, producing a shape called a hyperboloid. Objects are carried away from each other as space expands—the distance between A and B increases through time, and the further away an object is, the faster it appears to be moving.

object A at a time corresponding to the left-most circle, and arrives at object B at a time corresponding to the right-most circle, we see that the circumference of the circle has increased in the meantime and the distance travelled is therefore greater than would be the case if the universe was static (see Fig. 10). The objects A and B are not moving *through* space, they are being *carried apart* by the expansion of space. Covering this extra ground stretches the wavelength of the light so that it is redshifted. This is still a cosmological effect.

Because the objects may not actually be moving at all, turning a cosmological redshift into a velocity is not really very meaningful. But astronomers had by now long been in the habit of doing just this using the Doppler formula, and the resulting velocities provided a handy reference. The redshift (which today is given the symbol z) is related to the measured change in wavelength of the light $\Delta\lambda$ by $z = \Delta\lambda/\lambda$, where λ is the unshifted wavelength as we would measure it in the laboratory. A redshift implies a longer measured wavelength, such that $\Delta\lambda$ (and hence z) is positive, and the larger the redshift, the larger the value of z.

The Doppler formula relates z to the velocity (v) of the emitting object divided by the speed of light, $z = v/c$.[13] Lemaître reasoned that the more distant the spiral nebula the greater its cosmological redshift. Assuming a linear relationship therefore suggests $z = \kappa \times d$, where d is the distance and κ (Greek, kappa) is a simple constant of proportionality—it is the slope of the plot of z vs d. Combining this with the Doppler formula gives $v = c \times \kappa \times d$, or $v = H_0 \times d$, where H_0 ($= c \times \kappa$) is just the constant of proportionality multiplied by the speed of light. The reason for choosing the symbol H_0 will soon become apparent.

Lemaître now took the velocity data (in km/s) for 42 of the nebulae as listed by Strömberg in 1925, and combined these with the apparent magnitudes from Hubble's 1926 survey. He used Hubble's relation, $\log d = 0.2m + 4.04$, to estimate their distances (in millions of parsecs, or megaparsecs, Mpc). Giving the data for each nebula equal weight, he deduced $H_0 = v/d = 575$ km/s per Mpc. These units (which I will hereafter write as km/s/Mpc) are obviously derived simply from velocity in km/s divided by distance in Mpc. If we convert Mpc into km,* then the units of H_0 reduce to 'per second', /s, or s^{-1}.

Although Lemaître had not thought to provide a graph of velocity against distance, it was clear to him that the data were still rather scattered and unreliable and it was therefore difficult to be definitive. It seems he did not hold much store by his estimate of the value of H_0, accepting that it was possible to assume only that the speed, v, is proportional to the distance, d, and for astronomers to seek to avoid systematic error in their determination of the ratio v/d.

Alas, Lemaître chose to publish his work in French in a rather obscure journal *Annales de la Société Scientifique de Bruxelles*. Just like Friedmann's papers of 1922 and 1924, it had no impact. When some of the world's leading physicists gathered in Brussels for the fifth Solvay conference in October 1927, Lemaître, who was not invited to participate, nevertheless arranged to meet with Einstein. The conference was held at the Solvay Institute of Physiology in Parc Léopold, and Einstein and Lemaître talked as they strolled around the park. Einstein appeared to be

---

* 1 Mpc $= 3.086 \times 10^{19}$ km (see Appendix 2).

Photo 4 Einstein and Georges Lemaître in conversation: 'After some favourable technical remarks, [Einstein] concluded by saying that from the physical point of view [the expanding universe] seemed to him quite abominable'.

aware of Lemaître's paper 'that a friend had made him read'. Many years later Lemaître recalled Einstein's reaction: 'After some favourable technical remarks, he concluded by saying that from the physical point of view [the expanding universe] seemed to him quite abominable'.

Seeking to prolong their conversation, Lemaître secured an invitation to join Einstein on a taxi ride to the University of Brussels, where he was to visit the laboratory of Swiss physicist Auguste Piccard. 'In the taxi, I talked about the speeds of nebulae and I had the impression that Einstein was hardly aware of astronomical facts'.[14] Einstein was still not convinced. It seems likely that during their discussion Einstein informed Lemaître about the work of Friedmann.

The Hubble–Lemaître Law

Hubble finally turned his attention to the question of the speeds of spiral nebulae in early 1928. He was aware of the 'de Sitter effect' and may have discussed the possibility of a linear velocity–distance relationship with Robertson at the California Institute of Technology (Caltech). Robertson had published some speculations of his own that year, and later recalled discussing these with Hubble, though Hubble himself never referred to Robertson's efforts. It appears that Hubble met with Lemaître at the third meeting of the International Astronomical Union in Leiden in July 1928. Slipher had exhausted the reach of the 24-inch telescope at Lowell Observatory, and Hubble resolved to expand Slipher's data set using the 100-inch telescope at Mount Wilson.

Humason set to the task, and found that NGC 7619,* an elliptical nebula in the constellation Pegasus, exhibited a redshift corresponding to an unprecedented velocity of 3779 km/s. Hubble had access to redshift data for 46 nebulae, and he had distances for 24 of these. After correcting for the motion of the Sun, a plot of velocity against distance for these nebulae produced a graph that looks roughly linear (these are the filled circles and bold line in Fig. 14), implying $v = H_0 \times d$, which henceforth I will write as $v = H_0 d$. Combining the data into groups

---

* NGC stands for New General Catalogue (of Nebulae and Clusters of Stars), compiled in 1888 by Danish astronomer John Dreyer (see Appendix 1).

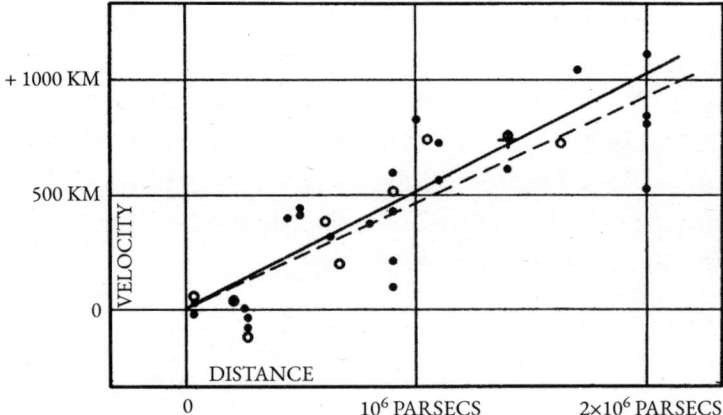

Fig. 14 The velocity–distance relation for spiral nebulae published by Hubble in 1929. See the main text for an explanation of the symbols and lines.

reduces the scatter somewhat (the open circles and dashed line in Fig. 14). Grouping the 22 nebulae for which individual distances could not be determined allowed Hubble to estimate a mean distance from the mean of their apparent magnitudes (the single cross in Fig. 14). The bold line corresponds to $H_0 = 530$ km/s/Mpc.

Hubble now used Humason's data for NGC 7619 to confirm that the linear relationship extends well beyond the dimensions shown in the figure. Correcting for solar motion increases its velocity to 3910 km/s, and assuming $H_0 = 500$ km/s/Mpc gives its distance as 7.8 Mpc. From the nebula's apparent magnitude of 11.8, Hubble deduced a distance modulus μ of 29.46, and hence an absolute magnitude of −17.65. 'A preliminary distance, derived independently from the cluster of which this nebula appears to be a member, is of the order of 7 [Mpc]'.[15]

Another way of looking at this is to divide the velocity of 3910 km/s by 7 Mpc, which gives $H_0 = 558$ km/s/Mpc, broadly consistent with the data shown in Fig. 14. In a companion paper published alongside Hubble's, Humason wrote: 'The high velocity for N.G.C. 7619 derived from these plates falls on the extrapolated line which expresses the relationship between line displacement and distance. These results suggest

Photos 5 and 6 Astronomers Milton Humason (left) and Edwin Hubble (right) established a linear relationship between the speeds of spiral nebulae and their distances.

an influence of distance upon the observed line shift—such as would be produced, for example, on de Sitter's theory, both by the apparent slowing-down of light vibrations with distance and by a real tendency of material bodies to scatter in space'.[16]

Hubble's paper was seen as significant, though not revelatory. Expectations within the astronomical community had been building for some time, and the data presented by Humason and Hubble in 1929 confirmed what many had already suspected. Hubble regarded himself as an astronomer first and foremost, and although in his paper he speculated (as Humason had) on the possibility of a connection between the linear speed–distance relationship and the de Sitter effect, he was broadly content to leave the interpretation of his data to others 'who are competent to discuss the matter with authority'.[17] He would continue to resist acknowledging any reference to an expanding universe until 1953, just a few months before his death.[18]

Hubble's results were discussed at a meeting of the Royal Astronomical Society on 10 January 1930. De Sitter confessed that he had struggled to accommodate a linear speed–distance relationship in his own 'solution B'. A discussion of Einstein's and de Sitter's universes ensued. Eddington expressed puzzlement that there are only two solutions, one static (Einstein), and the other non-static and expanding (de Sitter), 'but as there isn't any matter in it that does not matter'. De Sitter wondered what would happen if matter were inserted into his empty solution B: 'The difficulty in the investigation of this problem lies in the fact that it is not static'.[19] These comments were recorded in the proceedings of the meeting. When Lemaître read these, he fired off a letter to Eddington pointing out that he had completed such an investigation two years previously. Eddington had seen Lemaître's 1927 paper but, much to his embarrassment, he had forgotten all about it.[20]

Both Eddington and de Sitter now sought to set the historical record straight, and acknowledge the importance of Lemaître's contribution, and that of Friedmann before him. In March 1931 Lemaître's paper was republished in English translation in the *Monthly Notices of the Royal Astronomical Society*, but the section in which he had examined the speeds of nebulae and arrived at an estimate for H_0 was omitted. As the years passed the linear speed–distance relationship would become widely known as *Hubble's law*, governed by H_0, the Hubble constant. Conspiracy theories abounded. Was it possible that Lemaître's paper had been censored to ensure Hubble retained priority? Was Hubble himself the censor? Or was this simply the decision of the journal editor?

The answer was furnished by Israeli-American astrophysicist Mario Livio in November 2011. A search of the Royal Astronomical Society's archive unearthed a letter from Lemaître to the editor of the *Monthly Notices* dated 9 March 1931, from which it is apparent that Lemaître himself furnished the translation, and 'did not find advisable to reprint the provisional discussion of radial velocities [of the spiral nebulae] which is clearly of no actual interest'.[21] Lemaître had no desire to seek to establish priority for Hubble's law, and his 1927 speculations had been superseded by the new data and analysis from Humason and Hubble. This does not mean that Lemaître did not seek or deserve *recognition* for his pioneering work, and in 2018 the members of the International Astronomical Union voted overwhelmingly to rename Hubble's law as the *Hubble–Lemaître law*.[22]

The Einstein–de Sitter Universe

Einstein recanted his static universe in 1930, likely following discussions with Eddington, who had by then embraced Lemaître's expanding universe with some enthusiasm. He was therefore likely already a convert *before* visiting the Mount Wilson Observatory towards the end of that year, though he told journalists that the redshift data had 'smashed my old construction like a hammer blow'.[23] He published a short paper in April 1931 in which he accepted that the universe is expanding, and asked rhetorically 'whether one can account for the facts without the introduction of the [Λ]-term, which is in any case theoretically unsatisfactory'.[24]

Einstein had felt obliged to introduce the cosmological term in 1917 as a way to prevent his universe from evolving in time. If the universe is actually expanding, as the astronomical data now appeared to suggest, then there was surely no need to keep this term. With no small sense of relief, he now abandoned it, confessing that it was the biggest blunder he ever made in his life,[25] though as we will see, both Eddington and Lemaître argued that it should be retained.

Einstein also acknowledged that dynamic solutions of his gravitational field equations did not exclude the possibility of flat Euclidean space ($k = 0$) or negatively curved space ($k = -1$). The astronomical data pointed firmly in the direction of an expanding universe, but there was no observational evidence for anything other than flat space. Observations were not limited to the classroom: all the physical phenomena observed by astronomers could be understood by assuming the validity of Euclidean geometry. Having blundered with the cosmological term, it appears that Einstein was also now willing to abandon his insistence on a positively curved, closed universe.

Consequently, in the expanding universe that Einstein devised in collaboration with de Sitter in 1932, space is assumed to be flat and the cosmological term is gone.[26] It is easy to lose track of the assumptions inherent in these early models of the universe, so I've attempted to summarize their main features in Table 1 below.

The Einstein–de Sitter universe is assumed to contain just enough matter to ensure that its mutual gravitation applies a brake on the rate of expansion, such that after an infinite amount of time the expansion will cease. This would later be recognized as a 'critical' (or 'Goldilocks') density of matter, given the symbol ρ_c, required precisely to balance the expansion and ensure that space in the Einstein–de Sitter universe remains flat.

It will help in what follows to introduce the dimensionless parameter Ω (Greek omega), defined as $\Omega = \rho/\rho_c$, the ratio of the density of matter or mass–energy in the universe to the critical density. A universe with $\Omega > 1$ has a density of matter in excess of the critical density. Such a universe is therefore closed and space is positively curved. In

Table 1. Characteristics of early models of the universe

Author (Year)	Matter?	Static/Dynamic	Spatial Curvature	Λ-Term
Einstein (1917)	Yes	Static	$k = +1$	Yes
De Sitter (1917)	No	Static(?)	$k = +1$	Yes
Friedmann (1922)	Yes	Expanding/ Contracting	$k = +1$	Yes/No
Friedmann (1924)	Yes	Expanding/ Contracting	$k = -1$	Yes/No
Lemaître (1927)	Yes	Expanding	$k = +1$	Yes
Einstein–de Sitter (1932)	Yes	Expanding	$k = 0$	No

the Einstein–de Sitter model, $\Omega = 1$ and space is flat. A universe with $\Omega < 1$ is open and space is negatively curved.

In 1936, the English mathematician Arthur Walker independently rediscovered the spacetime metric that describes a space that is both homogeneous and isotropic, and this became subsequently known as the Robertson–Walker metric. When used in the field equations devised by Friedmann and rediscovered by Lemaître, the result is known as the Friedmann–Lemaître–Robertson–Walker (FLRW) metric.

The Einstein–de Sitter universe and the FLRW metric would become the basis for cosmological research for many decades.

The Age Paradox

In 1931, Humason and Hubble extended the speed/distance data set by a further 40 nebulae, with one nebula in the constellation Leo exhibiting a redshift corresponding to a velocity of 19,700 km/s. The new data extended the dimensions of the velocity–distance relationship from 2 Mpc (Fig. 14) to 35 Mpc, a factor of almost 20. There could now be no doubting the linearity of the Hubble–Lemaître law, with H_0 determined to be 558 km/s/Mpc.

But something was not right. In 1926 the US National Research Council established a committee on the 'Physics of the Earth', with a sub-committee on the 'Age of the Earth', whose purpose was to resolve significant discrepancies between different geological dating methods. Arthur Holmes, a British geologist and pioneer of the radiometric dating of minerals, was a founder member of the sub-committee. In its report, published in 1931, he presented a detailed review of the available radiometric data, from which he concluded: 'No more definite statement can be made at the present than that the age of the Earth exceeds 1460 million years, is probably not less than 1600 million years, and is probably much less than 3000 million years'.[27]

Theories of the composition and evolution of stars were also undergoing rapid development in this period (and we will return to this subject shortly), but were not yet sufficiently mature to provide a definitive estimate for the age of the Sun. Nevertheless, there were grounds for concluding that the Sun must be billions of years old, and at least as old as the Earth.

If the universe is expanding, this implies that it must have had an 'origin' in some distant past or, at least, a beginning to its present cycle. Assuming the expansion rate has not changed in the time since this beginning allows us to derive an estimate for the 'Hubble age' of the universe simply as the reciprocal of the Hubble constant. The reciprocal of $H_0 = 558$ km/s/Mpc is 1.75 billion years. But, in the Einstein–de Sitter universe, the rate of expansion slows with time due to the gravitational effects of the critical density of matter. Under these circumstances the age of the universe is then further reduced by a factor of two-thirds, to 1.17 billion years. Already in 1931 it seemed that the universe was, nonsensically, *younger* than the Sun and the Earth.

Either the models were incomplete and misleading, such that the universe is actually eternal and there was no beginning, or the Hubble constant (and the distance ladder on which its determination was based) was in error.

4

Divine Curves of Creation

Anyone encountering the expanding universe for the first time will instinctively want to ask the question: *why* is it expanding? This question was not lost on the pioneers of early relativistic cosmology, though at the time many steered clear of it as it wasn't obvious that this was a question with a scientific answer. Eddington had shown that the original Einstein static universe was unstable. He compared it with a pin balanced finely on its point. Any small disturbance of the delicate balance between gravity and the cosmological term would tip the pin over and spill the universe into instability. Perhaps the conversion of matter into radiation energy (according to $E = mc^2$) would be sufficient to cause instability, though it wasn't obvious that the result would necessarily be expansion.

Consequently, for both Eddington and Lemaître, Einstein's cosmological term and its associated constant remained essential ingredients in the description of our own universe. For Eddington, this was the 'hidden hand' driving expansion, and eradicating it from the theory—as Einstein and de Sitter had now done—was a step backwards as unthinkable as reverting to Newton's non-relativistic physics.[1] A positive cosmological term also offered a potential solution to the age paradox, as it competed with gravity and allowed the possibility that the rate of expansion had not been uniform in the past. If this rate had been much slower following whatever instability had triggered it, then a simple extrapolation backwards in time would inevitably underestimate the age of the universe.

But there was only so much extrapolation that Eddington was willing to countenance. Like Einstein, he favoured a universe without a beginning for the simple reason that he judged that any speculation about 'beginnings' was unscientific. 'Views as to the beginning of things lie almost beyond scientific argument. We cannot give scientific reasons why the world should have been created one way rather than another ... Since I cannot avoid introducing this question of a beginning, it has seemed to me that the most satisfactory theory would be one which made the beginning *not too unaesthetically abrupt*'. In other words, Eddington preferred a universe that persisted in an Einsteinian static or very near-static state, with 'no hurry for anything to begin to happen', until the pin inevitably succumbed to its inherent instability and toppled over. Stars and galaxies followed, setting the universe on its expansionist journey. To those who would beg to differ, Eddington would say: 'Have it your own way. And now let us get away from the Creation back to problems that we may possibly know something about'.[2]

The Fireworks Universe

Lemaître had no such qualms, and there's little doubt that Eddington was addressing these remarks to him. We might be tempted to jump to the conclusion that Lemaître the Catholic priest was in pursuit of a mathematical description of the theological concept of Creation, but there is plenty of evidence to suggest that he successfully compartmentalized these aspects of his life. He was principally concerned to draw satisfactory *scientific* conclusions from a theory that had been found to enjoy support from empirical data. And he believed he had an answer to Eddington's concerns: 'Sir Arthur Eddington states that [in an article published in *Nature* magazine], philosophically, the notion of a beginning of the present order of Nature is repugnant to him. I would rather be inclined to think that the present state of quantum theory suggests a beginning of the world very different from the present order of Nature'.

If we look at Lemaître's speculations in the context of scientific thinking and understanding of the time (1931), then I think we might agree that he was quite remarkably prescient. The neutron had not yet been discovered. Radioactivity was understood to involve the breakup of larger atoms variously into smaller atomic fragments, high-energy electrons (beta-particles), and high-energy photons (gamma rays). The origin of highly penetrating radiation, discovered in 1912 to be flooding the Earth's upper atmosphere (called 'cosmic rays' by Robert Millikan in the 1920s), was a mystery. Arguments about their nature bounced back-and-forth. Did they consist of highly energetic particles, as some believed, or high-energy photons, as others argued?

For closed solutions of Einstein's gravitational field equations, extrapolating an expanding universe backwards in time leads to a singularity—an unphysical point of infinite density. Lemaître sought to avoid this by proposing that, at the beginning, all the energy in the universe was compacted into a single, highly unstable atom. '... we could conceive the beginning of the universe in the form of a unique atom, the atomic weight of which is the total mass of the universe. This highly unstable atom would divide [into] smaller and smaller atoms by a kind of super-radioactive process'. But this is not just an origin of matter and radiation: 'If this suggestion is correct, the beginning of the world happened a little before the beginning of space and time. I think that such a beginning of the world is far enough from the present order of Nature to be not at all repugnant'. He subsequently estimated that the primeval atom would have had a radius of about 10^{14} centimetres (cm), or 1 billion km, a little larger than the orbital radius of Jupiter. Lemaître's primeval atom, containing the entire universe, fits comfortably inside the solar system.

There was still the tricky question of the 'moment' of creation of the universe, and the perennial philosophical conundrum of a 'first cause'. A priest might be content to invoke a creator God, but a scientist demands instead an explanation based on physical mechanism. Lemaître believed that such a mechanism is inherent in the quantum nature of atomic and sub-atomic particles. In the scientific revolution that took place in the mid–late 1920s, there had emerged an understanding of quantum physics based on the foundations of indivisible quanta, wave–particle duality, a very specific kind of quantum probability, and uncertainty. The heart of the very smallest components of matter and radiation suffer a peculiar arrhythmia, beating uncertainly, the outcomes of physical events predictable only in terms of probabilities.

The connection between cause and effect, taken for granted in the mechanics of Newton, is considerably weakened in the mechanics of the quantum. Here was Lemaître's escape clause. A quantum fluctuation, governed by uncertainty, is, like a spontaneous radioactive decay, an effect without apparent or direct cause. The decay of Lemaître's 'primeval atom' is spontaneous and probabilistic. Many different kinds of universe may result: '... the whole story of the world need not have been written down in the first quantum like a song on the disc of a phonograph ... the story it has to tell may be written step by step'.[3]

Some inkling of how Lemaître sought to reconcile his inner priest and his inner scientist may be gleaned from his final sentence: 'I think that everyone who believes in a supreme being supporting every being and every acting, believe also that God is essentially hidden and may be glad to see how present physics provides a veil hiding the creation'. He thought better of it, and chose to cross out this sentence before submitting his short manuscript to *Nature*.[4]

This was no more than a bare-bones set of ideas, but over the next couple of years Lemaître gathered these into a formal model, which he referred to as a 'fireworks theory'. His key challenge was to use the theory to predict something that could potentially be observed and thus hold the promise of providing supporting evidence. One possibility was that the explosive release of energy resulting from the initial disintegration of the primeval atom might have left a tell-tale signature, a relic of the tumultuous beginning of the universe: the 'ashes and smoke of bright but very rapid fireworks'. These, he thought, were the cosmic rays, which the fireworks theory demanded should be high-energy particles rather than photons.

The fireworks universe would have expanded rapidly as the primeval atom was broken down into stars, before settling into a period of stagnation, or at least of relative calm, expanding little and with characteristics not too dissimilar from Einstein's original static universe, as gravity vied with the cosmological term. There followed a period of rapid, accelerated expansion as the term containing the cosmological constant became a dominant force. 'It is doubtless in this third period that we find ourselves today, and the acceleration of space which followed the period of slow expansion could well be responsible for the separation of the stars into extra-galactic nebulae'.[5]

Vacuum Energy

In Lemaître's fireworks theory, the cosmological constant took on a different significance compared with Einstein's original intentions. Einstein had configured the cosmological term on the left-hand (spacetime) side of the field equations, as a kind of anti-gravitational pressure which serves to offset the extent of spacetime curvature caused by all the matter on the right-hand (mass–energy) side. In 1933 Lemaître suggested that: 'Everything happens as though the energy *in vacuo* would be different from zero'.[6] In other words, the cosmological term represents an energy of empty space, also known as *vacuum energy*.

We tend to think of empty space or the vacuum as devoid of all content. What Lemaître had deduced was that space throughout the universe might itself possess an intrinsic energy. This is very different from the more familiar energy associated with matter and radiation. Its precise origin and nature are unknown, and Lemaître did not volunteer an explanation. He estimated the density of matter in the universe to be of the order of 10^{-30} grams per cubic centimetre (g/cm^3), and a density of vacuum energy corresponding to $\Lambda c^2/4\pi G$, or $\sim 10^{-27}$ g/cm^3. There is a thousand times greater density of vacuum energy in Lemaître's fireworks universe than there is of matter.

Although Lemaître himself did not explicitly take this step, we are perfectly entitled to gather all the mass–energy terms on the right-hand side of the field equations and so move the (negative) cosmological term from left to right, where it becomes positive. If we do this we get:

$$\text{Spacetime curvature} = 8\pi\,G \times \text{Mass-energy} + \text{Cosmological term}$$

Compare this with the earlier expression. The cosmological term now represents a positive contribution to the total mass–energy of the universe, consisting of matter, radiation, and vacuum energy. In anticipation of what is to come, we should note here that in Lemaître's fireworks universe, the dimensionless density parameter Ω now has two contributions, one from the density of matter or mass–energy, Ω_M, and one from the density of vacuum energy, which we will denote Ω_Λ: $\Omega = \Omega_M + \Omega_\Lambda$.

In the Einstein–de Sitter universe, there is no cosmological term and $\Omega = \Omega_M$. Einstein, now a convert to the 'abominable' expanding universe, was surprisingly enthusiastic about the fireworks theory, although he remained determinedly unenthusiastic about the cosmological term. But the astronomical community and the relativistic cosmologists remained reserved. For many the very notion of a 'beginning' to the universe was counter-cultural. This was all very interesting, perhaps, but there was no evidence to support what some judged to be 'speculation run mad'.[7]

The evidence was restricted to the here and now, and all we could conclude from it is that the universe is expanding at a rate measured by the Hubble constant. Concluding anything else was a matter of philosophical taste. 'Not until the empirical resources are exhausted, need we pass on to the dreamy realms of speculation', wrote Hubble, in the very final sentence of his book *The Realm of the Nebulae*, published in 1936.[8] And, although the fireworks theory offered a potential solution to the age paradox (later, in 1958, Lemaître suggested that, based on his model, the age of the universe lies between 20 and 60 billion years), many continued to be deeply troubled by it.

The community moved on, eager to embrace the authority of the Einstein–de Sitter universe and willing to assign the cosmological term and its constant to the small catalogue of rare Einstein blunders. The fireworks theory was quietly ignored.

Stellar Nucleosynthesis

Although still some years away, the discovery that would greatly ease (though not fully resolve) the age paradox would arise from a growing appreciation of the composition and evolutionary history of stars. In science, as in life, sometimes when you're searching for something the best strategy is to look for something else. We will therefore briefly turn our attentions away from grand theories of the cosmos to the physics of the bright things that sit within it.

The H–R diagram offered insights on the evolution of stars, but questions about the source of a star's energy remained. The dominant hypothesis involved the conversion of gravitational energy into heat and light as a star contracts. Some judged this explanation inadequate (it suggested that the Sun should be about 20 million years old, and so much younger than the Earth). Eddington, for one, was forthright: 'Only the inertia of tradition keeps the contraction hypothesis alive—or rather, not alive, but an unburied corpse'. In an address to the British Association in Cardiff in August 1920, Eddington went on to make the following remarkable speculation:[9]

> The nucleus of the helium atom, for example, consists of four hydrogen atoms bound with two electrons. But ... the mass of the helium atom is less than the sum of the masses of the four hydrogen atoms which enter into it ... There is a loss of mass in the synthesis amounting to about 1 part in 120 ... If 5 per cent of a star's mass consists initially of hydrogen atoms, which are gradually being combined to form more complex elements, the total heat liberated [from the transmutation of mass into energy] will more than suffice for our demands, and we need look no further for the source of a star's energy.

Once again, this suggestion has to be seen in historical context. The compositions of stars were unknown, and many (such as Russell) argued in favour of the hypothesis that the Sun and Earth are of similar elemental composition. In 1925 Cecilia Payne would argue in her Radcliffe College PhD thesis that stars are composed mostly of hydrogen and helium, but when Russell disputed this she added the sentence 'Although hydrogen and helium are manifestly very abundant in stellar atmospheres the actual values derived from the estimates of marginal appearance are regarded as spurious'.[10] Russell would realize his mistake and credit Payne for her discovery four years later, though the scientific community would continue to credit Russell for many years.*

To provide sufficient energy to power the Sun for ∼15 billion years, Eddington had assumed that *all* the hydrogen would be consumed in nuclear fusion reactions in the stellar furnace at the core. This was most unlikely, however, and recognizing that stars are composed predominantly of hydrogen rendered this assumption unnecessary.

In addition to this we should note—once again—that the neutron had yet to be discovered. The helium nucleus actually consists of two positively charged protons and two electrically neutral neutrons, which is why only two negatively charged electrons are required to balance the positive charge in the neutral helium atom. But Eddington had correctly identified the principal source of the energy of a star about the size of the Sun. In what is now called the proton–proton chain, four protons (four hydrogen nuclei) are fused together in a sequence of nuclear reactions to produce a single helium nucleus, requiring that two protons are converted into two neutrons in a process that is the reverse of beta-radioactive decay. Eddington was also almost right about the mass difference—known as the *mass defect*—which is about 0.7% of the combined mass of the four protons (about 1 part in 140). This 'missing' mass is converted into energy according to $E = mc^2$.

---

* Payne's struggles and ultimate triumph against prejudice and a reactionary academic establishment are recounted in Stella Feehily's play *The Lightest Element*. In 1976, Payne (by then Payne-Gaposhkin) gave the Henry Norris Russell Lecture, awarded by the American Astronomical Society. Karma.

The details of stellar nucleosynthesis were worked out in the late 1930s, by George Gamow, Hungarian physicist Edward Teller, and German theorist Carl Friedrich von Weizsäcker. But it was German theorist Hans Bethe who began properly to unravel the mystery of energy production in stars in a paper published in 1939, in which he correctly identified the initial reaction involving the fusion of two protons to form a deuterium nucleus, a heavy isotope of hydrogen consisting of a proton and neutron.[11] He went on to describe what is now known as branch II of the proton–proton (p–p) chain, which contributes about 16% of the Sun's luminosity. He also identified an important catalytic cycle involving carbon, nitrogen, and oxygen atoms which also converts four protons into a helium nucleus and which operates in larger stars. Branch I of the p-p chain, which converts four protons more directly into a helium nucleus, as Eddington had speculated, contributes about 83% of the Sun's luminosity.

Eddington's Valve

A star like the Sun enters a long, relatively quiescent period on the main sequence, starting out towards the bottom right of the H–R diagram and ascending towards the top left. During this period the force of gravity which threatens to contract the star is balanced by the internal pressure resulting from the flow of heat released by nuclear reactions in the core. This internal pressure supports the star against collapse and holds it steady, in *hydrostatic equilibrium*. Whilst this might be helpful in maintaining life on a nearby planet like Earth, it was clear that not all stars behave in this way. Variable stars like the Cepheids are not quiescent: they pulsate.

Eddington speculated that the Cepheids are not in hydrostatic equilibrium. As a Cepheid star contracts, its core temperature and pressure rapidly increase, eventually overcompensating for the inward pull of gravity and so rapidly expanding again. The expansion over-shoots and the cycle of contraction and expansion repeats. It's as though the star is breathing in and out. Such an oscillation would be expected to be damp-ened and eventually come to a halt, just as a swinging pendulum will slow to a stop. To prevent this from happening, Eddington proposed a kind of valve mechanism. According to this, compressing the layers of gas makes them more opaque, blocking the outward flow of radiation, trapping the heat which then builds up. As the layers expand their opacity decreases, and the heat is released. The star behaves like a heat engine.

But compressing layers within a star more typically causes opacity to decrease, not increase, the opposite of what Eddington's valve requires. The solution was discovered in the 1960s. When layers of gas containing partially ionized helium atoms (He^+, with just a single electron removed) are compressed, the extra energy strips off the remaining electron to pro-duce doubly ionized helium, He^{2+}, which is more opaque. Compression of these layers then leads to increased opacity, as required.

The location of the helium ionization zone depends on the star's temperature. Cooler stars with temperatures of about 4500 K* have ionization zones close to their centres. Hot stars with temperatures of 7500 K have their ionization zones close to their surfaces. Only stars of intermediate temperatures produce pulsation. Cepheids are therefore typically high-mass stars that have begun to evolve away from the main sequence into a region of the H–R diagram known as the 'instability strip'.

Stellar Populations

As further spectra were gathered, the pattern of the H–R diagram for the population of nearby stars became more refined. The mirror-7 shape was found to be too simplistic, and it was realized that there exists a gap between the main sequence and the red giant branch—a general absence of stars of spectral type A5 to G of magnitude between −3 and +1. This

---

* The kelvin (K) is the unit of the absolute temperature scale, related to the more familiar celsius (°C) scale by the simple relation 0 °C = 273.15 K. There is no temperature below absolute zero, 0 K.

became known as the Hertzsprung gap. We now understand that as a star exhausts its supply of hydrogen, producing an inert core of helium, it starts to burn hydrogen in its outer layers. Its density falls as the star swells to form a red giant. This happens very quickly—within a thousand years or so—which in the big scheme of the universe is no time at all. The Hertzsprung gap arises simply because the probability of observing a star making this transition within any given population of stars is fleetingly small.

It was also soon realized that the pattern of the H–R diagram obtained for our stellar neighbours is not characteristic of all stars. Studies of stars in globular clusters revealed very different patterns, with sequences that begin at K-type and move *downwards* from right to left, splitting around spectral type G into two branches, one continuing downwards towards F and the second sweeping horizontally towards A and B, with significant numbers of stars sitting in the Hertzsprung gap. These stars were clearly following very different evolutionary paths.

In 1926, the Dutch astronomer Jan Oort had argued that, among nearby stars, there appeared to be two types depending on the speeds with which the stars are moving. The H–R diagram of fast-moving stars shows virtually no examples of type-O or B, and a much lower proportion of red giants. In other words, the fast-moving stars sit in H–R diagrams that look rather like those of the globular clusters.

It was German astronomer Walter Baade who drew these threads together. Baade had moved to the US to take up a permanent position at the Mount Wilson observatory in 1931. He had fully intended to apply for American citizenship but disliked bureaucracy. When he mislaid his application documents in a house move, he gave up on the effort. Consequently, at the start of the second world war he was classed as an enemy alien and confined to Pasadena. Fortunately, the terms of his confinement allowed him continued access to the observatory.

As observatory staff committed themselves to essential war work, Baade was left with lots of time on the 100-inch Hooker telescope. The wartime black-out of Los Angeles produced near-ideal observing conditions, and in 1942 Baade was able to use red-sensitive photographic

Fig. 15 The Andromeda Nebula (galaxy), M31, showing the two 'early-stage' elliptical galaxies studied by Baade in 1942.

plates to resolve stars in the central region of the Andromeda Nebula (catalogued as NGC 224 and Messier 31, or M31*) and in its satellites M32 and NGC 205 (see Fig. 15). This kind of resolution should in principle have been possible only with the 200-inch Hale telescope, then under construction at the Palomar observatory, on Palomar Mountain in northern San Diego County. Baade's achievement was 'a pure steal; a steal from the prestige of the yet uncompleted 200-inch telescope which was considered the only telescope capable of the job'.[12]

The nebulae M32 and NGC 205 had been classified by Hubble as 'early-type'. These are elliptical galaxies, at the time thought to be at an

---

* This is an earlier catalogue of astronomical objects compiled by French astronomer Charles Messier, published in a preliminary version in 1774.

early stage of galactic evolution on their way towards spiral structures. This was later found to be incorrect. The 'early-type' elliptical nebulae are in fact older than many spiral galaxies. Their origin is still something of a mystery. Baade noticed that the spectra of stars in M32 and NGC 205 formed patterns that looked very much like the H–R diagrams of stars in the globular clusters. 'This leads to the further conclusion that the stellar populations of the galaxies fall into two distinct groups, one represented by the well-known H–R diagram of the stars in our solar neighbourhood (the slow-moving stars) the other by that of the globular clusters'. These two distinct patterns are shown in Fig. 16.

Baade called the first group type I, the second type II:[13]

Although the evidence presented in the preceding discussion is still very fragmentary, there can be no doubt that, in dealing with galaxies, we have

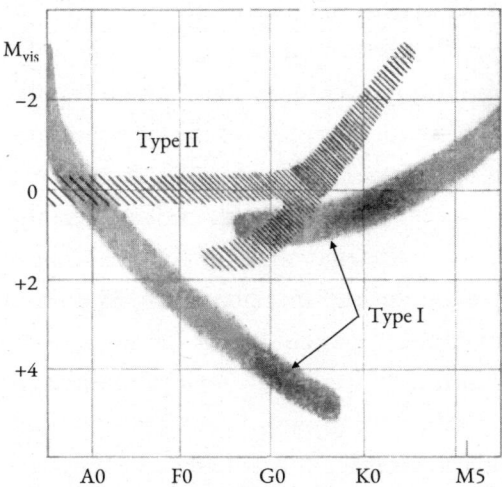

Fig. 16 Baade compared the H–R diagram derived from the spectra of nearby stars (the shaded areas—note the Hertzsprung gap between the main sequence and red giant branch) with the very different diagram derived from the spectra of stars in globular clusters and 'early-type' nebulae such as M32 and NGC 205 (the hatched areas). He concluded that these differences reflect the existence of different star populations—type I and II.

to distinguish two types of stellar populations, one which is represented
by the ordinary H–R diagram (type I), the other by the H–R diagram of
the globular clusters (type II). Characteristic of the first type are highly
luminous O- and B-type stars and open clusters; of the second, globular
clusters and short-period Cepheids. Early-type nebulae [such as M32 and
NGC 205] seem to have populations of pure type II. Both types coexist,
although differentiated by their spatial arrangement, in the intermediate
spirals like the Andromeda nebula and our own galaxy.

The origin of the difference wasn't obvious to Baade, but we now
know that 'Population II' stars (Baade's type II) are older than 'Population I' (Baade's type I). The Population II stars are characteristically of
lower 'metallicity'—they possess smaller proportions of elements heavier than hydrogen and helium compared with Population I—and they
are found in globular clusters and the central bulges of Andromeda and
the Milky Way. The H–R diagrams of Population II stars do not feature
the bright O- and B-type stars because examples of these have long since
disappeared from the population.

Understanding the relation between the two population types and
their places in the evolutionary story of the universe would require some
further ingredients. With his Mount Wilson colleague Fritz Zwicky,
Baade had in 1934 already identified the key step that would prove
to be essential to our understanding of their relationship. These were
the relatively infrequent but spectacular bursts of energy, sometimes
visible to the naked eye, representing the release of 'a considerable fraction of the star's mass'.[14] Baade and Zwicky referred to these events as
supernovae.

When a Population II star goes supernova, all the elements in the peri-
odic table up to uranium are formed. As the resulting gas and dust from
exploded stars are gathered together once more by gravity to form a new
generation of younger Population I stars, their composition is necessarily
contaminated by heavier elements. As the cloud responsible for a Population I star's formation swirls and condenses, some of the heavier elements
are flung out to form rocky planets, and the resulting star (like the Sun)
has a higher metallicity.

Alpher, Bethe, Gamow

Gamow secured his PhD at the University of Leningrad* in 1928, three years after Friedmann's untimely death. Although he had initially developed interests in relativity, under Krutkov's guidance he turned his attentions to quantum mechanics. He spent the summer of 1928 in Göttingen. Subsequent periods at Niels Bohr's Institute for Theoretical Physics in Copenhagen and with Ernest Rutherford at the Cavendish Laboratory in Cambridge completed his transformation into a nuclear physicist. He returned to the Soviet Union to join the State Radium Institute in Leningrad in 1929, and two years later was elected to the Academy of Sciences of the USSR, as one of its youngest members. He helped to initiate the building of the Institute's first particle accelerator in 1932.

The Soviet Union experienced a series of crises in 1932–33, and life in Stalinist Russia became intolerable as the authorities introduced ever more severe sanctions. Like many other physicists Gamow was denied permission to attend scientific conferences outside the country. He decided to defect with his new wife Lyubov, also a physicist, whom he affectionately nicknamed 'Rho' (the Greek letter ρ), and sought alternative escape routes to the West. Their attempt to flee in a kayak 170 miles across the Black Sea to Turkey was foiled by bad weather: 'We ... spent three days in a storm on the Black Sea and finally were thrown ashore, 60 miles away from the place where we started'. They had taken with them enough food for five days, two bottles of brandy, five dollars, and a borrowed English driving licence.[15]

Much to his surprise, in 1933 he was delegated to attend the Solvay Conference in Brussels on the structure and properties of the atomic nucleus. He later discovered that Bohr and French physicist Paul Langevin had worked behind the scenes with the Soviet authorities to facilitate his invitation. Despite personal reassurances received in a meeting at the Kremlin with Vyacheslav Molotov, Chairman of the Council of People's Commissars, the Soviet authorities initially declined

* Petrograd was renamed Leningrad following Lenin's death in 1924.

to issue a passport for his wife. He dug his heels in and the authorities eventually relented. With further help from Marie Curie in Paris, they both defected. Gamow was appointed to a professorship at George Washington University in Washington, DC in 1934, and became a US citizen in 1940.

Despite Gamow's reputation as a nuclear physicist, he grew rather bored of the subject. He had chosen to work on nuclear problems in 1928 because it was 'a corner where nobody was doing anything'. For rather obvious reasons, post-war nuclear physics 'blew up into a big thing'. Gamow preferred to be a pioneer, in unexplored territory: 'I would rather go into these mountains [in Colorado] than in California, where they have a hot dog stand on the top of each mountain'.[16] He chose instead to work on nuclear astrophysics and cosmology. At this time cosmology was barely recognized among his peers even as a scientific subject and it was hardly popular. Here he could at least be sure that, on reaching the top of a mountain representing some stubborn problem in cosmology, he wouldn't find lots of other physicists already buying hot dogs.

Gamow set himself the task of working out the details of *primordial nucleosynthesis*, the production of chemical elements in the earliest moments of an expanding universe. It was now widely accepted that, as Cecilia Payne had deduced, stars are formed predominantly of hydrogen (about 76%) and helium (about 24%), with the heavier elements rapidly diminishing in abundance with increasing atomic weight. Gamow realized that the element abundances would have been determined by the *competition* between the expansion and cooling of the early universe and the speeds of the nuclear reactions involved. If we attempt to fill a bath with the plug removed, after a time the amount (abundance) of water in it will depend on the competition between the rate of filling from the tap and the rate of draining through the plughole. Fill the bath too slowly (expand the universe too quickly) and we won't gain enough of the heavier elements. Fill the bath too quickly (expand the universe too slowly) and we end up with too much.

Photo 7 In 1948, George Gamow set himself the task of working out the details of primordial nucleosynthesis.

He was completely unaware of Lemaître's fireworks theory from 20 years previously, and chose as a starting point a highly compressed gas consisting entirely of neutrons (which, by this time, had been discovered). He did not speculate on how this gas might have formed. Unlike the proton, a free neutron is radioactive, with a half-life (the time required for half of an initial number of neutrons to decay) then measured to be about 12 minutes. It decays into a proton and a high-speed electron (a beta-particle). At the high temperatures assumed to prevail in the universe's earliest moments, Gamow imagined that neutrons would fuse with the protons formed by radioactive decay to produce deuterium nuclei. Building up the elements would then proceed by further neutron capture and beta-radioactivity, turning some of the accumulated neutrons into protons.

Gamow worked on these problems alone but in 1948 was joined by his PhD student Ralph Alpher. With some assumptions about the initial density of neutrons in a universe dominated by matter, they made what appeared to be a passable prediction of the element abundance curve. On submitting a paper describing their calculations to the journal *Physical Review*, Gamow chose to add Bethe's name to the list of authors, flagged as 'in absentia'. Bethe had not been involved in the work but Gamow had a reputation as a prankster. The possibilities afforded by a paper authored by Alpher, Bethe, and Gamow had captured his imagination. Inevitably, it became known as the alpha-beta-gamma paper. The journal editor checked with Bethe, who had no objection to being included as the paper might well be right, and the 'in absentia' was removed. The paper was published on 1 April 1948.[17]

Alpher was unimpressed by Gamow's antics, although one consequence of the stunt was that Gamow was able to persuade Bethe to serve on the committee that was to examine Alpher's dissertation. As co-author of their paper, Bethe would be unable to speak against it: 'Ain't that smart!'.[18] Such was the publicity surrounding Alpher's thesis defence that it was moved to a large hall at George Washington University to accommodate the 300 people and reporters who turned up.

The Cosmic Background

The alpha-beta-gamma paper contained significant errors, some corrected in Alpher's thesis and in later publications by Gamow and by Alpher and Robert Herman, who had studied cosmology under Robertson. Gamow joked that he had tried to persuade Herman to change his name to Delter.

Gamow developed a different model based on an early universe dominated by hot radiation instead of matter and submitted a short note to the journal *Nature*. He sent a copy of the manuscript to Alpher. Alpher and Herman had been thinking along similar lines and, when they understood that Gamow's calculations were seriously in error, they hastily informed him in a telegram. Gamow judged that it was now too late to withdraw his paper, and instead urged Alpher and Herman to submit a note correcting his error for publication in the same journal.[19] Alpher and Herman's correction contains a rather prophetic prediction, based on a much more sophisticated model subsequently published in detail in *Physical Review* (by coincidence, on 1 April 1949). It's worth taking a little time to examine this.

Gamow had understood that the density of radiation in an early universe that was dominated by radiation would fall more rapidly than the density of matter as the universe expanded. This means that there would be a period when the densities of matter and radiation become equal, after which the universe is dominated by its matter content. The density of matter (ρ_M) in today's universe had been estimated (by Hubble, and others) to be about 10^{-30} g/cm^3. Alpher and Herman estimated ρ_M to be much higher (10^{-6} g/cm^3) in the early universe, with temperatures of about 10^8 to 10^{10} K, in order to allow for the possibility of nucleosynthesis. For simplicity, Alpher and Herman chose to set the density of radiation (ρ_r) to the value 1 g/cm^3 at a temperature of 6×10^8 K, at which neutron capture required for building up new elements becomes possible. According to their model, this assumption fixes ρ_r in today's universe at 10^{-32} g/cm^3. With the bounds set up this way, the equations predict how the early universe evolves to the present day as it expands and cools. The results are shown in Fig. 17.

This figure shows how four different parameters develop in time (t) from the earliest moments of the universe. These are the length or distance parameter, L, a measure of the distance to an object (such as a galaxy) at a specific moment in time, relative to today's distances such that L increases from 1 as the universe expands forwards from today and decreases as we wind backwards. The densities ρ_M and ρ_r decline (the matter and radiation become more diluted) as the universe expands, and the temperature, T, declines as well. Because the changes of scale are so

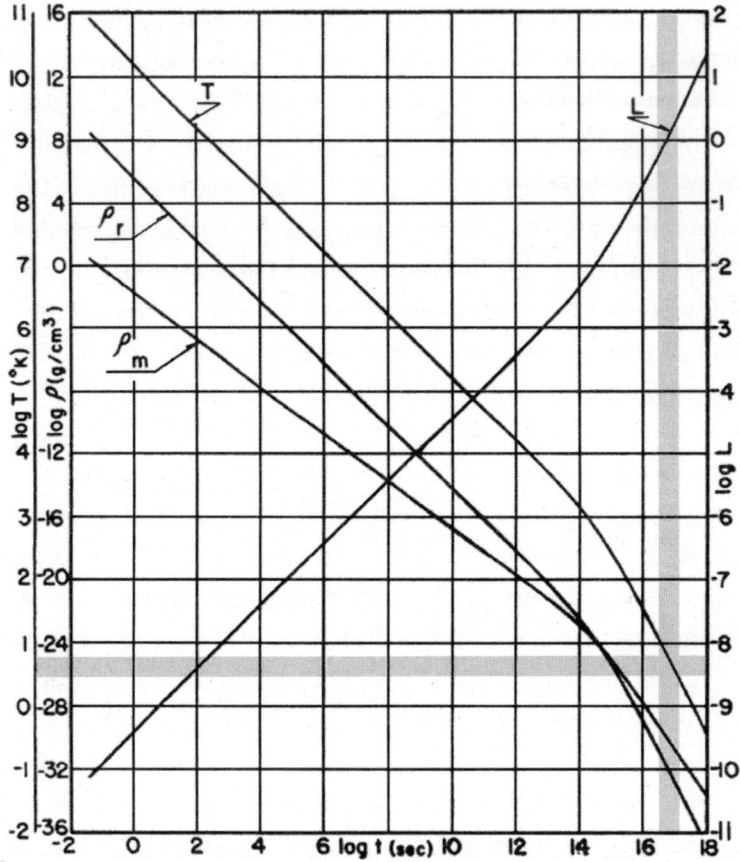

Fig. 17 In this 'divine curve' Alpher and Herman predicted that the background temperature of today's universe, 1.8 billion years ($\log t = 16.8$, $\log L = 0$) after its origin, would be about 5 K ($\log T = 0.7$).

large, the logarithms of these parameters are plotted. Gamow called such plots 'divine curves', and used one of them as part of the letterhead for some new personal stationery.[20]

In Fig. 17 the time axis runs from less than 0.1 seconds to almost 32 billion years. Assuming a Hubble age of 1.8 billion years fixes 'today' at about 57×10^{15} s, or $\log t = 16.8$. At this point in time, by definition $L = 1$ and $\log L = 0$. The calculations suggest a temperature of just 5 K ($\log T = 0.7$). 'This mean temperature for the universe is to be interpreted as the background temperature which would result from the universal expansion alone'.[21] This was the prediction that Alpher and Herman had published in their earlier note to *Nature*. Obviously, matter concentrated into stars and galaxies is much hotter than this background, so the implication is that the 5 K temperature is derived largely from the radiation, the relic of a much hotter, denser, period of the early universe.

We would come to know this as the *cosmic background radiation*.

This was a prediction that depended sensitively on the assumed densities. A lower matter density of 1.8×10^{-4} g/cm^3 in the early universe brought the cross-over point forward and pushed the present-day temperature down to 1 K. There was much uncertainty. Alpher and Herman were also uncertain about the extent to which stellar radiation might mask the cosmic background. Perhaps if they had simply translated the temperature into a spectrum—a black body with a temperature of 5 K emits predominantly microwave and infrared radiation—someone might have been persuaded to look for it.*

But the odds were stacked against them. Estimates of the background temperature (including separate estimates suggested by Gamow) varied from 1 K to 50 K. The age paradox remained unresolved. Later efforts to develop more detailed descriptions of primordial nucleosynthesis uncovered a significant problem, called the 'mass gap'. Whilst the abundances of hydrogen and helium were well accounted for, the building up of heavier elements by neutron capture couldn't bridge the gaps between the stable isotopes of helium and lithium, ^{4}He and ^{6}Li, and of another stable isotope ^{7}Li and beryllium, ^{9}Be. There are simply no stable isotopes with 5 or 8 'nucleons'. In other words, there are no stable nuclei with 2 protons and 3 neutrons (^{5}He) or 3 protons and 2 neutrons (^{5}Li) or with 3 protons and 5 neutrons (^{8}Li) or 4 protons and 4 neutrons (^{8}Be). These isotopes decay very rapidly, so do not build up sufficient quantities to capture further neutrons. This means that element building would be

---

* In 1940–41 the Canadian astrophysicist Andrew McKellar had suggested that an unidentified spectral line detected in interstellar space might be due to a transition corresponding to a temperature of 2.7 K. See Kragh, *Conceptions of Cosmos*, p. 180.

expected to grind to a halt long before heavier elements such as carbon or oxygen could be formed.

Alpher and Herman may also have suffered some guilt by association. Although Gamow could be an incredibly creative scientist, as Herman admitted: '... there were people who were just put off by George. So that may be an element in how Alpher and I might have been viewed'.[22]

There was also by now an important rival cosmology.

5

Parameters of the Universe

Young British theorist Fred Hoyle and Austrians Herman Bondi and Thomas Gold at Cambridge University had no problem with the notion of an expanding universe. But they perceived the age paradox to be a serious flaw that couldn't be fixed in any satisfactory way using Einstein's cosmological term. Bondi and Gold set out their position in a paper published in 1948. Any attempt to extrapolate backwards in time to an early, hot, dense state of the universe is based on the implicit assumption that the physical laws and constants as measured in the laboratory today remain valid in the distant past. But this implies that physical laws determined in the laboratory are somehow independent of the structure of the universe. A hot, dense universe would look very different. Is it not reasonable to suppose that different laws and constants might prevail in it? Circularity can be avoided by taking a different view. The physical laws and constants are unchanging because the universe is actually in a stable, self-perpetuating state:[1]

> We regard the reasons for pursuing this possibility as very compelling, for it is only in such a universe that there is any basis for the assumption that the laws of physics are constant; and without such an assumption our knowledge, derived virtually at one instant of time, must be quite inadequate for an interpretation of the universe and the dependence of its laws on its structure, and hence inadequate for any extrapolation into the future or the past.

They called it the *perfect cosmological principle*: the universe is not only homogeneous and isotropic, it is also stationary in its large-scale appearance as well as its physical laws. 'We do not claim that this principle must be true, but we say that if it does not hold, one's choice of the variability of the physical laws becomes so wide that cosmology is no longer a science'.

To maintain a constant density of matter as spacetime expands, such a 'steady-state' universe demands that new matter be continually and spontaneously created. In a companion paper published in the same journal, Hoyle demonstrated that by supplementing Einstein's gravitational field equations with a 'C-field', or creation-field, responsible for continuous matter creation, it is possible to produce a version of the de Sitter universe that is both stationary and expanding, without recourse to a cosmological term.

As matter is carried apart by expansion, new matter is created to take its place. Although it was not possible to be definitive about the nature of the particles that were being created: 'Neutron creation appears to be the most likely possibility. Subsequent [radioactive] disintegrations might be expected to supply the hydrogen required by astrophysics. Moreover, the electrical neutrality of the universe would be guaranteed'.[2] Later estimates suggested a rate of creation of new matter of about one atom per year in a volume of space equal to St Paul's Cathedral in London.

It was an uncomfortable choice. Although the steady-state hypothesis attracted some favourable attention in Britain, most astronomers and cosmologists rejected the idea. A steady-state universe implies that new galaxies are forming to replace those that simply expand beyond the horizon at which objects are being carried away faster than their light can reach back to us. In contrast, the Mount Wilson astronomers were convinced that all galaxies are the same age.[3]

We will return to the steady-state universe in due course, but it is worth noting that arguments in its favour did have one important (if unintended) side-effect: it gave a memorable name to the cosmology it sought to displace. In a BBC radio broadcast on 28 March 1949, Hoyle declared:[4]

> On scientific grounds this big bang hypothesis is much the less palatable of the two. For it is an irrational process that cannot be described in scientific terms... On philosophical grounds too I cannot see any good reasons for preferring the big bang idea. Indeed it seems to me... a distinctly unsatisfactory notion, since it puts the basic assumption out of sight where it can never be challenged by a direct appeal to observation.

Without any kind of empirical observation to distinguish between them, Gamow, Alpher, and Herman had no choice but to acknowledge the existence of a rival. '... whenever we lectured individually, separately or together, we always presented these views of an evolutionary universe in the "big bang" sense, and also the steady-state point of view'.[5]

It didn't help the cause of cosmology as a science that the only basis for choosing between the Big Bang and steady-state models appeared to be theological. In an address to the Pontifical Academy of Sciences in November 1951, Pope Pius XII declared that Lemaître's Big Bang cosmology provided scientific confirmation of the Christian doctrine of Creation. Lemaître hurried to the Vatican and—it seems—was successful in getting the Pope to accept that a scientific hypothesis or theory is always provisional and so cannot serve as proof for an article of religious faith. It was just as well. Pius XII was due to address the participants attending the next meeting of the International Astronomical Union, scheduled for Sunday, 7 September 1952, at Castel Gandolfo, his summer residence.

Baade's Correction

By 1939, the structure and dimensions of the Milky Way galaxy were broadly fixed and agreed by astronomers. Fig. 18 provides a contemporary representation.[6] Seen edge-on, the galaxy exhibits a central bulge

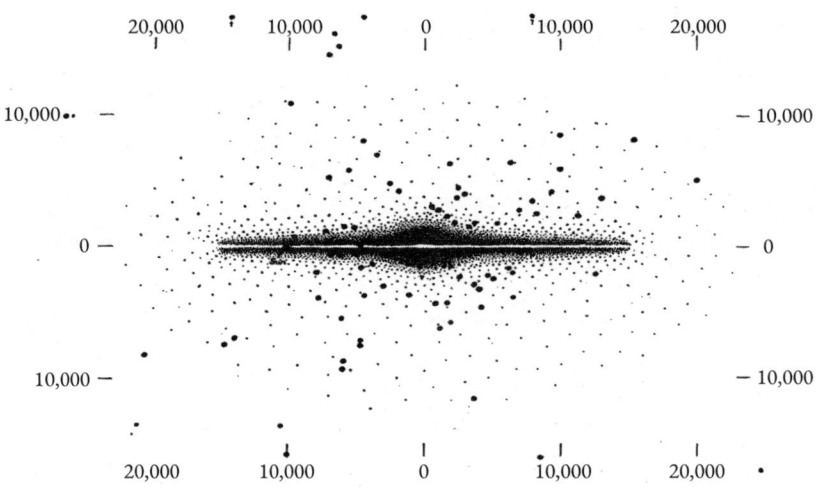

Fig. 18 By 1939, the structure and dimensions of the Milky Way galaxy were broadly agreed by astronomers. This diagram is fairly typical.

with a high density of stars which flattens to a disc as we move out along the spiral arms. The galaxy has a diameter of about 30,000 parsecs, and a disc thickness varying from about 1000 to 2000 parsecs. The Sun lies about two-thirds of the way from the centre (10,000 parsecs), shown on the left in Fig. 18. Dotted above and below the plane at distances up to 10,000 parsecs are scattered stars and the globular clusters, arranged in a roughly spherical distribution.

This structure is repeated in other spiral galaxies. When Hubble investigated the globular clusters of the Andromeda Nebula in 1931, he was puzzled to discover that they appeared about 1.5 magnitudes dimmer than their equivalents in the Milky Way. This was odd because, to all intents and purposes, Andromeda appeared to be a spiral galaxy much like our own, and there was no good reason to suspect that its globular clusters would be different. Baade recalled numerous discussions about the possible reason, 'especially on cloudy winter nights on Mount Wilson', when there was nothing better to do.

The answer lay in Baade's conclusions from his work in 1942. He realized that the Milky Way Cepheids and the Cepheids in the Magellanic Clouds that Leavitt had used to establish the original period–luminosity relation all belong to Population I. Long-period Cepheids and other variable stars in the globular clusters all belong to Population II, 'and there was no a priori reason to expect that two Cepheids of the same period, the one a member of Population I, the other of Population II, should have the same luminosity'.[7] In fact, there are two different period–luminosity relations, one for each population, and Shapley's calibration had not distinguished between them.

Hubble had determined the distance to Andromeda using Population I Cepheids, and the estimates of the luminosities of its globular clusters were based on this distance. In contrast, the distances and luminosities of the Milky Way globular clusters were all derived from older Population II stars. Baade estimated that to correct for this, it was necessary to decrease the zero point of the relation for Population I Cepheids by 1.5 magnitudes. In other words, the Population I Cepheids are all intrinsically *brighter* than first thought.

Confirmation came with the aid of the new 200-inch Hale telescope at Palomar. According to Shapley's calibration, cluster variables known as RR Lyrae stars should have just been visible in Andromeda with the new telescope, but they were not. The archetype is RR Lyrae in the constellation Lyra. These are Population II stars, commonly found in globular clusters, and first identified as 'cluster-type' variable stars by Williamina Fleming in 1901. But other Population II stars known to be 1.5 magnitudes brighter than the RR Lyrae stars *were* visible. Baade concluded that Population II stars must all be generally 1.5 magnitudes dimmer than Population I.

Baade wrote to David Thackeray, a former Mount Wilson colleague who had moved to the Radcliffe Observatory in Pretoria, South Africa, to suggest he use the newly-installed 74-inch reflecting telescope to search for RR Lyrae stars in the Magellanic Clouds. Thackeray confirmed Baade's suspicions.[8]

What did this mean for distance estimates? Baade and Thackeray presented their results at a meeting of Commission 28 of the IAU, part of the larger meeting held in Rome in September 1952.[9] The Commission had been set up to examine extra-galactic nebulae. Hubble was its President, and its membership included Baade, Bondi, Humason, Lemaître, Lundmark, Oort, Shapley, Slipher, and Zwicky, among others. Hoyle served as secretary.

A 1.5 magnitude decrease implies that a star is actually $(2.512)^{1.5} \sim 4$ times brighter than previously thought. The 'classic' (Population I) Cepheids were therefore four times brighter and (from the inverse-square law) twice as far away.* Hubble's estimate for the distance to the Andromeda Nebula instantly doubled to 2 million light-years (and it now appeared that Andromeda is bigger than the Milky Way). Hubble's constant was halved, to about 280 km/s/Mpc. And, as Hoyle noted in the meeting minutes, 'Above all, Hubble's characteristic time scale for the Universe must now be increased from about 1.8 [billion] years to about 3.6 [billion] years'.[10]

Not surprisingly, Shapley demanded to know the details of Baade's correction, which Baade duly supplied. Shapley's zero point had been based on Population I Cepheids, and was 1.5 magnitudes too large, making them dimmer than they actually are. When applied to Population II Cepheids, which were in their turn 1.5 magnitudes dimmer than anyone had realized, the error cancelled out. The same was true for Shapley's calibration applied to the RR Lyrae variables, explaining why independent estimates were in good agreement. Compounding the errors in this way meant that they had gone unnoticed for more than 30 years.

Baade's correction was not yet sufficient fully to resolve the age paradox, but it was a huge step in the right direction. And it suggested that the basis for the distance calibration warranted much more attention than it had so far received.

---

* If you're not used to dealing routinely with stellar magnitudes, this doesn't sound right. But remember, magnitudes work backwards. A star judged to be 10 parsecs distant has a distance modulus $\mu = 0$. But suppose we now discover that its absolute magnitude has been underestimated, and is in fact 1.5 magnitudes brighter (more negative). Its apparent magnitude hasn't changed, so from $\mu = m - M$ we deduce a new distance modulus $\mu = 1.5$. From $d = 10^{(\mu+5)/5}$, we get $d = 10^{1.3}$ or $d \sim 20$ parsecs. The star is twice as far away.

H II Regions

Astronomy at Mount Wilson was supported by the Carnegie Foundation, established by the Scottish-American industrialist Andrew Carnegie in 1911. But Hale had appealed to the Rockefeller Foundation, established a few years later by the oil magnate John D. Rockefeller, for funding for the Hale telescope at Palomar, about 150 miles to the south. Rather than hand the funds to a rival foundation, the Rockefellers instead gave the money to Caltech. The Mount Wilson astronomers became (unpaid) Caltech professors, and although the institutions remained separate, they were run jointly as the Mount Wilson and Palomar Observatories.

Hubble was already aware of Baade's correction in 1951, when he set out the cosmological research programme for the new Hale telescope. There were four main themes: the large-scale distribution of galaxies; the law of redshifts; the cosmic distance scale; and cosmological theory.[11] Cosmology was badly in need of a proper empirical basis and Hubble, following a suggestion of Eddington, hoped that astronomical data might help to determine the nature of the curvature of the universe. If this could be done, it might be possible to distinguish between the different universes—or 'world models'—that the theorists had wrought.

Working with Humason and Nicholas Mayall at the Lick Observatory, over a period of 20 years Hubble had built up a catalogue of redshifts for 620 nebulae observed at Mount Wilson and Palomar, including 26 clusters of nebulae, and 300 nebulae observed at Lick. These figures included 114 nebulae studied by both groups. The largest redshifts, found for clusters of faint nebulae, corresponded to speeds of 60,000 km/s, one-fifth of the speed of light.

It was another Hubble *tour de force*, but he would not live to see its publication. Already in poor health from two heart attacks, in September 1953 he suffered a stroke, from which he did not recover. There was no

Photo 8 Allan Sandage was asked to pick up where Hubble had left off.

funeral and his widow, Grace, never revealed whether or not he had been cremated or where he had been buried.

Baade and Humason were approaching retirement, and the challenge of building on Hubble's legacy fell to a young astronomer from Iowa who had studied for his PhD at Caltech under Baade's supervision. 'It was all laid out, and I was the only one left after Hubble died'. His name was Allan Sandage. Humason and Mayall were ready to publish, and Humason now asked Sandage to pick up where Hubble had left off, analysing the relationship between the magnitudes and redshifts of the nebulae. 'If you were assistant to Dante and then Dante died, and then you had in your possession the whole of the *Divine Comedy*, what would you do? What actually would you do?' Sandage trembled at the size of the task, and the Oxford brogues he was being asked to fill, but it was an extraordinary opportunity. 'If I didn't grab that opportunity I would be nuts'.[12]

The Humason, Mayall, Sandage paper was published in 1956. The data were fairly unequivocal: although there was still much scatter, the linear Hubble–Lemaître law extended out to the largest redshifts. But Sandage had become painfully aware of the rickety nature of the distance ladder that Hubble had erected 20 years earlier. Hubble had used two sources of data. The first was the mean absolute magnitudes of the brightest stars in the so-called 'Local Group' of galaxies, which includes the Milky Way's nearest galactic neighbours,* calibrated using Cepheid variables. The second was a relation between redshift and apparent magnitude for the brightest stars in more distant nebulae. Baade's correction for the Cepheids already implied that Hubble had underestimated the distances to the Local Group. Sandage now found problems with the data for the distant nebulae as well.

Hubble had taken advantage of bright 'knots' of light in the spiral arms of distant nebulae by assuming these were ordinary (non-variable) stars. Closer inspection of photographic negatives taken using different colour filters revealed that they were not stars, but rather bright clouds of ionized hydrogen, called H II regions, masking stars about 2 magnitudes fainter. The naming convention is as follows: clouds of neutral hydrogen atoms (H) are called H I; clouds of ionized hydrogen (H^+) are H II; and clouds of molecular hydrogen (H_2) are called H_2 (not to be confused with H II). The H II regions are now known to be associated with star formation (the Orion Nebula is a large H II region), and are characteristically red in colour due to a strong emission line associated with neutral atomic hydrogen, called the H_α line (see Fig. 4). The (unshifted) wavelength of the H_α line is 656 nm. It is the brightest visible line in the H-atom emission spectrum.

Sandage highlighted the problem in an appendix to the 1956 paper.[13] The difference in brightness due to H II regions was revealed in photographs of a spiral arm of the nebula NGC 4321 in the Virgo cluster

* The term Local Group was coined by Hubble in *The Realm of the Nebulae*. It includes the Milky Way, Andromeda (M31), M32, NGC 205, the Large and Small Magellanic Clouds, Triangulum (M33), and many others.

(a)   (b)

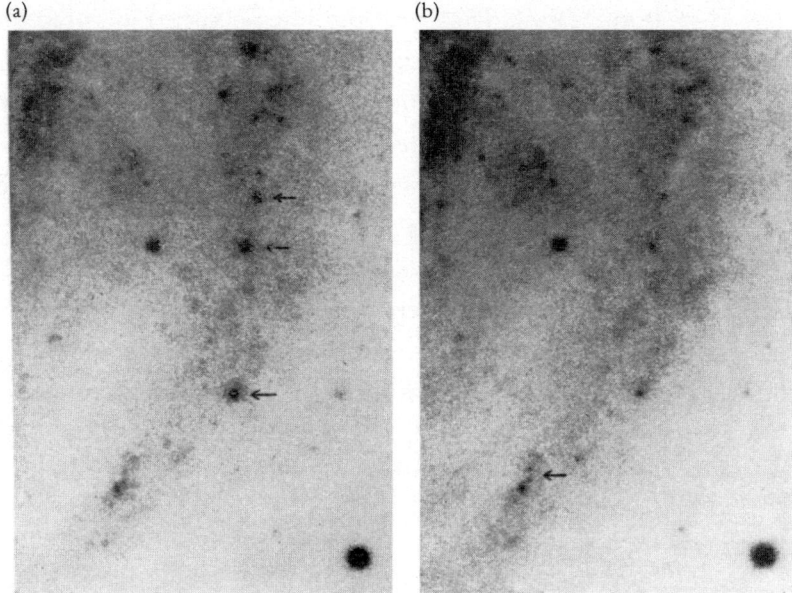

Fig. 19 In 1936 Hubble had identified what he thought were bright stars in NGC 4321 and had used these to help calibrate his distance ladder. These are marked with arrows in (a), taken with a filter combination which isolates light between 630 and 670 nm and which includes the H_α line. Contrast this with (b), taken with a filter combination which isolates light between 520 and 630 nm (and so excludes the H_α line).

(see Fig. 19), taken using the 200-inch telescope. Compare Fig. 19(a), taken with a filter combination which isolates light between 630 and 670 nm (red) and which includes the wavelength of the H_α line, to Fig. 19(b), taken with a filter combination which isolates light between 520 and 630 nm (green-to-orange/red), and so excludes the H_α line. With red light from the H II regions blocked, fainter stars are revealed beneath the clouds. Hubble had clearly overestimated the apparent magnitudes (m) of these stars. For a given absolute magnitude (M), this meant that the distance modulus $\mu = m - M$ had been underestimated (remember, overestimating m makes it smaller or more negative than it really is). NGC 4321, and hence the Virgo cluster, lay further away.

Hubble had suggested a distance modulus for the Virgo cluster of 26.7, implying a distance of about 2.19 Mpc. The mean redshift of nebulae in the cluster had been determined to correspond to a speed of 1136 km/s. From $H_0 = v/d$ it follows that Hubble's 1936 estimate of the distance suggests $H_0 \sim 519$ km/s/Mpc, reasonably consistent with the value 558 km/s/Mpc from 1929. But Sandage now deduced that NGC 4321 is a little more than 2 magnitudes fainter, suggesting a revised distance modulus of 29.05, and hence a distance of 6.457 Mpc. This implies $H_0 \sim 176$ km/s/Mpc. Sandage thought this was likely to be accurate to within 20%, and settled for $H_0 \sim 180$ km/s/Mpc, giving a Hubble age of about 5.6 billion years.

Sandage went much further just two years later.[14] By correcting the distances to the galaxies in the Local Group and by adopting different assumptions about the extent to which Hubble had confused stars with H II regions in more distant galaxies he was able to deduce values for H_0 ranging from 125 to 55 km/s/Mpc. He suggested $H_0 \sim 75$ km/s/Mpc, extending the Hubble age to about 13 billion years, with a possible uncertainty of a factor of 2.

Parameters of the Universe

The value of H_0 provided a critically important connection with the various model universes that the theorists had devised, and Sandage was determined to improve the accuracy and precision of its measurement. H_0 is a measure of the rate of expansion of the universe as judged 'today', from the disposition and the redshifts/speeds of galaxies as we observe them. The 'big bang' universes devised by Lemaître, Einstein–de Sitter, and by Gamow, Alpher, and Herman imply that the rate of expansion may have changed over time. This means that the Hubble 'constant' is not necessarily constant throughout the lifetime of the universe. It is more correct to call it the Hubble parameter and I can now confess that the subscript '0' in H_0 signifies its value as we measure it 'now', a time usually signified as t_0, the present time as measured from the Big Bang.

Think of it this way. Suppose we plot the proper distance between two galaxies against time (equivalent to the parameter L in Alpher and

Herman's divine curve shown in Fig. 17). If the rate of expansion had been uniform throughout the past and the same rate was extended into the future, this plot would be a straight line with a slope of H_0. If, however, the rate of expansion changes over time the result is a curve. A closed universe means the rate of expansion slows as the mass–energy of the universe applies its gravitational brake. The plot of distance (or L) vs time turns downwards, and the value of H_0 is then the slope of the *tangent* to this curve drawn at t_0. The same applies for the Einstein–de Sitter universe, in which there is just enough density of matter to halt the expansion and close the universe after an infinite amount of time. One consequence is that the age of such a universe is no longer given simply as the reciprocal of H_0, and the 'Hubble age' is therefore not necessarily equal to the age of the universe.

Hubble's ambitions for the Hale telescope had extended to cosmological theory, and Sandage now plunged himself into the mathematics, aided by his wife Mary. It took him five years, 'But at the end of that period I had some insight, some intuition'.[15] He judged that it might be possible to use astronomical data from the Hale telescope to distinguish between the different model universes using just two parameters. One was H_0, the other was the *deceleration parameter*, q_0, a measure of the extent to which the rate of expansion is slowing, as measured now.

The two are interconnected, and it will be worth our while taking the time to understand how. Imagine we could travel back into the past to some time t and measure the proper distance between two galaxies. We'll call this distance d_t. We expect that, due to subsequent expansion, we would measure this distance to be larger today: d_0 would be greater than d_t. We can generalize the ratio d_t/d_0 as a_t/a_0, where a_t is a dimensionless *scale factor*. In an expanding universe, the ratio a_t/a_0 is less than 1 at times earlier than t_0, and greater than 1 at later times (and it is this ratio that is equivalent to L in Fig. 17).

If the expansion rate as measured by H_0 had indeed been constant over the history of the universe, then we would conclude that

$L = a_t/a_0 = 1 + H_0 (t - t_0)$. We can see how this works. At times t less than t_0 (the past), the term $H_0 (t - t_0)$ is negative and L is less than 1. At times t greater than t_0 (the future), the term is positive and L is greater than 1. At $t = t_0$, $L = 1$.

But what if we now assume that the rate of expansion of the universe has changed over time? We don't have to assume a specific cosmological model to work out what might happen. Instead, we can adopt a trick that mathematicians use routinely, which is to 'expand' the expression for L as a *Taylor series*. In principle, the series has an infinite number of terms ascending in a power series, $(t - t_0)$, $(t - t_0)^2$, $(t - t_0)^3$, and so on. In practice, if the relation deviates only slightly from a straight line, we can get away with stopping at the term in $(t - t_0)^2$:

$$L = \frac{a_t}{a_0} = 1 + H_0 (t - t_0) - \frac{1}{2} q_0 H_0^2 (t - t_0)^2 + \ldots$$

In this expression H_0 is the rate of change of a_0 with time, divided by a_0. The term q_0 is the rate of change of the rate of change of a_0 with time, divided by $a_0 H_0^2$.* Note that the term in which q_0 appears is negative, so a positive value for q_0 has the effect of slowing the rate of expansion—at any given time L is smaller than it would be if q_0 were zero. This is why q_0 is a *deceleration* parameter.

Different cosmologies will give different sets of relationships for H_0 and q_0. In a matter-dominated universe described by the Friedmann equations and the FLRW metric, we find that $q_0 = \frac{1}{2}\Omega_M - \Omega_\Lambda$. Recall that $\Omega = \rho/\rho_c$, the ratio of the density of mass–energy to the critical density required for a spatially flat universe. Recall further that $\Omega = \Omega_M + \Omega_\Lambda$, where Ω_M is the density parameter for the

---

* We don't have to stop at the term in $(t - t_0)^2$. The next term can be written as $\frac{1}{6} j_0 H_0^3 (t - t_0)^3$, where j_0 is the 'jerk factor' or 'cosmic jerk'. This name is based on the classical jerk (or sometimes jolt) vector—the rate of change of acceleration—and has a literal interpretation. A sudden change in acceleration whilst driving a car can cause serious injuries such as whiplash. Further parameters in such an expansion are interpreted somewhat less literally. They are called 'snap', 'crackle', and 'pop'.

mass–energy derived from matter and Ω_Λ is the density parameter of vacuum energy associated with the cosmological constant, Λ.

We now have all we need to make a few comparisons, just as Sandage did in 1961 (see Fig. 20). For an open, Friedmann universe with no deceleration ($q_0 = 0$), $\Omega = \Omega_M < 1$, space is negatively curved, the universe is 'open', and the plot of scale factor vs time is a straight line. In this case the Hubble age is equal to the age of the universe. In the flat Einstein–de Sitter universe, the cosmological constant Λ is assumed to be zero, $\Omega = \Omega_M = 1$, and so $q_0 = \frac{1}{2}$. The universe is 'closed' but the scale factor reduces back to zero only after an infinite time. Notice that extrapolating backwards using the tangent to this line shows that the Hubble age overestimates the age of the universe (in fact, t_0 is equal to two-thirds of the Hubble age). In a closed Friedmann universe, $\Omega = \Omega_M > 1$, q_0

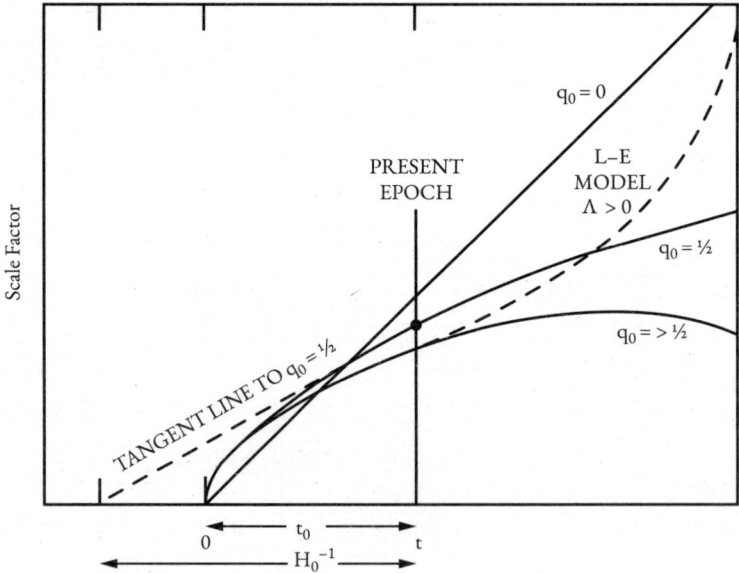

Fig. 20 Sandage compared the various model universes on the basis of two parameters—the Hubble constant H_0 and the deceleration parameter q_0. He hoped that extension of the Hubble diagram to larger and larger redshifts would allow astronomers to determine which model most closely describes our own universe.

is greater than ½, space is positively curved and the universe will collapse in on itself after a finite time.

The plot for what Sandage calls a generalized 'Lemaître–Eddington' universe (denoted L–E in Fig. 20) is also worthy of note. In this model, the cosmological constant is retained. The precise nature of the plot then depends on the balance between the densities of matter and vacuum energy, but in general the deceleration parameter approaches the limit $q_0 = -1$. A negative deceleration indicates that the expansion of such a universe is actually accelerating—galaxies are carried apart at faster and faster rates—driven by the density of vacuum energy. It so happens that the deceleration parameter for a steady-state universe is also -1, making it impossible to choose between these particular models based on these two parameters alone.

Nevertheless, the implications were profound. Hubble had initially tried to determine the curvature of the universe by counting the numbers of galaxies observed at different distances. Sandage had joined Hubble in this effort, but he now concluded that a better way of addressing the question would be simply to extend the Hubble diagram—the plot of redshift vs magnitude or distance—and look for deviations from straight-line behaviour.

The reason for such deviations is fairly straightforward. The light from distant objects takes a finite time to reach us on Earth, which means that when it arrives we see the object not as it is now, but as it was when the light was emitted. The more distant the object, the greater the 'look-back' time (see Appendix 3). The different curves shown in Fig. 20 are schematic, but they illustrate a general principle: different models of the universe suggest different 'histories' for the rate of expansion. If it were possible to look back far enough using very high-redshift objects, it might be possible to discern changes to this rate, which would show up in a Hubble diagram as deviations from a straight line.

This was asking a lot. The largest redshifts reported in the Humason–Mayall–Sandage paper were of the order of $z = 0.2$, corresponding to a time when the universe was about 83% of its current size. Looking back further would require even more distant, and so even fainter, objects. Sandage estimated that redshifts of $z = 2$ would be required to discriminate between the different models. Even with the Hale telescope, the prospects did not seem very promising.

But things were about to change.

The Quasar Revolution

In 1933, Karl Jansky at Bell Telephone Laboratories in Holmdel, New Jersey, finally tracked down the source of the constant hiss of radio noise he could detect with his purpose-built 100-foot directional antenna. It appeared and disappeared on an almost-but-not-quite daily basis, and was coming from the constellation Sagittarius, towards the centre of the Milky Way. His discovery generated little interest beyond a few newspaper headlines.

But after the second world war English physicist (and former radio ham) Martin Ryle at Cambridge University adapted some war surplus radar equipment to build a radio interferometer, and so helped to establish the new field of radio astronomy. There were objects in the sky that were emitting more than visible light, and extending the spectrum to include radio waves gave astronomers another means to explore the universe. By the end of the 1950s, Ryle and his colleagues had identified 450 bright radio sources, which they had collected in the *Third Cambridge Catalogue*. The objects, thought to be 'radio stars', were prefixed 3C (for third catalogue). Ryle thought the objects were extragalactic, distributed like galaxies with many more faint objects than bright ones. But the long wavelengths of the radio waves meant that they could not be easily pinpointed in the sky.

In late 1948 the American radio astronomer John Bolton and his colleagues had successfully identified radio sources in Centaurus A and Virgo A with galaxies NGC 5128 and M87, respectively. He decided to write to the astronomers at Palomar and ask for help tracking down the identities of other radio sources. The choice was Baade or Rudolph Minkowski (nephew of Hermann Minkowski) who, working with Baade, had earlier classified supernovae as Type I and Type II, differentiated by their spectra. Bolton had no basis on which to make a choice, so he tossed a coin. He duly wrote to Minkowski, only to receive a reply from Baade. He therefore addressed his reply to Baade, only to receive a further reply from Minkowski. Both were fascinated by the challenge he had posed.[16]

When Baade retired, Minkowski continued the search, though he was himself near to retirement. In his last run on the Hale telescope in 1960, he spent two consecutive nights gathering light from the direction of the radio source 3C 295, sufficient to determine its optical spectrum. What he found was truly astonishing. The object (whatever it was) exhibited a redshift $z = 0.46$, a speed of 138,000 km/s, nearly half the speed of light. If this was a cosmological redshift (at that time a big if), then its light was emitted when the universe was 68% of its current size.

Sandage got in on the act. He found that the brightness of light from another radio source, 3C 48, varied over a period of just a few weeks. This is behaviour characteristic of individual stars, not whole galaxies. It also had a rather bizarre emission spectrum.

Minkowski's successor, Maarten Schmidt, had studied at Leiden Observatory under Oort and had moved to Caltech in 1959. In 1963, he used the Hale telescope to measure the optical spectrum of another radio source, 3C 273, which had been associated with a small blue 13th-magnitude star, bright enough to be visible through a small commercial telescope. Schmidt didn't know what to make of the spectrum. But then he noticed a pattern in the spacings of the lines. These were emission lines characteristic of hydrogen atoms, but they were in the wrong part of the spectrum. Their positions could only be reconciled by assuming a redshift $z = 0.158$, corresponding to a speed of 47,400 km/s. This explained the bizarre spectrum of 3C 48, too. Its lines were found to exhibit a redshift $z = 0.37$, corresponding to a speed of 111,000 km/s.

This was something completely new: objects with the apparent size of stars, but with redshifts more characteristic of entire galaxies. And if the Hubble–Lemaître law applied to these objects, then their redshifts suggested vast distances and long look-back times. That they could be seen at all suggested that they were the brightest objects in the universe. If this was indeed a cosmological redshift, Schmidt estimated that 3C 273 must be about 500 Mpc distant. But it could not be much more than a kiloparsec in diameter, about 3% of the diameter of the Milky Way.[17]

These objects became known as 'quasi-stellar radio sources', which quickly became inappropriate once it was understood that only a few emit radio waves. An alternative was 'quasi-stellar objects', shortened to 'quasars'.[18] Although this name was received rather unenthusiastically by the astronomy community, it stuck. Today we understand that these are 'active galactic nuclei'—galaxies consisting of many millions or billions of stars surrounded by a disc of gas and powered by a supermassive black hole at the centre. The prodigious quantity of light they produce results from heating the infalling gas as it is drawn into the black hole.

This was another opportunity that Sandage couldn't pass up. Redshift records tumbled with every new discovery—in 1965 *Time* magazine reported that the quasar 3C 9 exhibited a recession speed of 149,000 miles per second (about 240,000 km/s, 80% of the speed of light). This corresponds to a redshift $z = 2.019$,* and a scale factor one-third of the present size of the universe. The article quotes Sandage suggesting that these might be 'very distant, superbright galaxies reaching more than halfway to the horizon of the universe'.[19]

Sandage applied the counting method to the quasars and concluded that the deceleration parameter q_0 is of the order of 1, implying a closed universe. In a Caltech press release timed with the publication of Sandage's paper in the *Astrophysical Journal* in May 1965, Sandage

---

* As the recession speeds approach the speed of light we can no longer use the approximation $z \cong v/c$. Instead, we must use the full Doppler formula, which can be rearranged to give $v/c = (z^2 + 2z)/(z^2 + 2z + 2)$. A redshift of 2 implies $v/c = 0.8$, or $v = 80\%$ of the speed of light, 240,000 km/s.

was quoted as saying: 'The clues indicate that our universe is a finite, closed system originating in a "big bang", that the expanding universe is slowing down, and that it probably pulsates perhaps once every 82 billion years'.[20] A scientific prediction that the universe will end, even in 69 billion years' time, attracted considerable media attention. Walter Sullivan, considered the dean of American science writers, penned an article about it in the *New York Times*.[21] Alas, in his hurry to realize his dream, Sandage had made some unjustified assumptions and some mistakes. Some of the quasars he had included in his analysis were actually white dwarf stars. His conclusions were premature.

It would take Sandage some years to fix the problems his haste had helped create, but the science of cosmology was in any case overtaken by other events. Two short papers, also submitted to the *Astrophysical Journal* in May 1965 (but published in July) would change our understanding of our universe forever.

Reinventing the Big Bang

Like Sandage, Princeton University physicist Robert Dicke also favoured an oscillating universe. The field equations for an expanding universe could be quite reasonably—and safely—tuned backwards in time but would eventually run up against the theoretically ugly singularity, at 'time zero'. An oscillating universe which periodically bounced from a collapsing, fiery 'big crunch' to an expanding hot Big Bang could in principle avoid such mathematical unpleasantness. All evidence of the previous universe—all the heavy elements that were now thought to be manufactured in supernova explosions—would be stripped back to matter and radiation at their most elementary, then thought to be protons, neutrons, electrons, strange new particles called mesons, and photons. Temperatures of the order of 10^{10} K would be required.

Dicke realized that radiation at this temperature would cool as the universe expanded, and might persist today as a cold remnant of the hot fireball—a fossil, almost—the tell-tale sign of a more exciting time in the young universe. 'I remember making a crude estimate that it [the temperature of the radiation today] would be about 45 degrees [K] or something like that. If it was much more than that, we'd get too much energy. The energy density would be too high'.[22]

Dicke was almost uniquely placed to do something about his realization. He had worked on radar during the second world war, and in 1946 he had invented a microwave radiometer, called a Dicke radiometer, designed to detect microwave radiation with a specific wavelength of 1.25 cm (12.5 mm).* A black body emits radiation across a broad range of wavelengths, and its brightness at these different wavelengths can be calculated from the radiation law devised by Planck in 1900. As its temperature falls, Planck's law determines that the peak brightness declines and shifts to longer wavelengths. A black body with a temperature of 45 K would be expected to have peak brightness at around 6.5 mm, corresponding to a frequency of 46 billion cycles per second, or 46 gigahertz (GHz), in the so-called 'Q-band' of the microwave spectrum. This is problematic, as water vapour in the atmosphere begins to absorb microwaves at around 40 GHz, and would be expected to block any background radiation coming from all points in the sky. But the black-body spectrum extends out to much longer wavelengths and, although it is substantially less bright, Dicke believed that the cosmic background could be detected at wavelengths of 1 cm or 3 cm.

He mentioned this idea to the small gravity research group gathered in his laboratory at Princeton in the summer of 1964. The group included experimentalists Peter Roll and David Wilkinson, and theorist James Peebles. 'Wouldn't it be fun if someone looked for this radiation?' Dicke asked.[23] Neither Roll nor Wilkinson had any experience in radio astronomy or microwave electronics, but they agreed to build a radiometer atop

---

* Microwave ovens are designed to operate at longer wavelengths, between 25 and 38 mm, and work by exciting the water molecules present in foodstuffs.

Princeton's Palmer Physical Laboratory, and learn as they went. Dicke turned to Peebles: 'Why don't you go and think about the theory?'

Peebles went home and thought about it. The inclination of most scientists would be to start at the library and search through the scientific literature, but Peebles had a blind spot. 'I was never strong on the literature ... It's so much more fun to think things through on your own than to read someone else's paper'.[24] Peebles reinvented big bang cosmology for himself. He used it to calculate the primordial abundance of helium, and to predict a temperature for the cosmic background radiation of 10 K.

In February 1965, Peebles was invited to present a colloquium on his work at the Applied Physics Laboratory at Johns Hopkins University in Maryland. He asked Wilkinson if it would be okay for him to mention the experiment that Wilkinson was building with Roll. Wilkinson gave his blessing. He figured that anybody in the audience motivated to develop their own experiment would be unable catch up with the Princeton group. In the audience was Kenneth Turner, a former Dicke PhD student and good friend of Peebles, who was now working as a post doc with Bernard Burke at the Department of Terrestrial Magnetism of the Carnegie Institute in Washington, DC. Greatly intrigued by the experiment, he subsequently described it to Burke.

A few weeks later Peebles submitted a paper to the *Physical Review* titled 'Cosmology, Cosmic Blackbody Radiation and the Cosmic Helium Abundance'. It was sent to Herman for review, who shared it with Alpher so that they could review it jointly. The paper itself was fine, though in their judgement it contained little that was new. But Alpher and Herman were appalled that Peebles had failed to refer to any of their earlier work from 1949. In their report they gave a list of references for him to check, but his revised manuscript still came up short '... he still had not, in our view, really looked at what we had done', explained Alpher.[25] They recommended that the paper not be published.

Dicke, Peebles, Wilkinson, and Roll were in the habit of gathering in Dicke's office every Tuesday lunchtime to discuss progress. One Tuesday (likely in March or April 1965), their meeting was interrupted by a phone call: 'That often happened; Dicke was a famous guy so people called him all the time'. Wilkinson later thought the call was short—about 5 minutes. Roll remembers differently, and thought the call lasted 30–40 minutes. Dicke listened intently and muttered a few remarks before putting the phone down. 'Well boys', he said, 'We've been scooped'.[26]

Excess Antenna Temperature

About 30 miles due east of Princeton, at the Bell Telephone Laboratories' Holmdel research facility in New Jersey, history had repeated itself. Like Jansky before them, radio astronomers Arno Penzias and Robert Wilson had been puzzling for some time over the source of annoying interference, this time occurring in the microwave region at a wavelength of 7.35 cm (4.08 GHz). It appeared to be coming from all directions in the sky. Despite their best efforts, this interference just wouldn't go away.

They were using a 20-foot horn antenna that had been built as part of NASA's Echo I satellite communications project. It was subsequently adapted to receive signals from the Telstar 1 satellite, launched with great fanfare in July 1962, and which was used successfully to relay the first television pictures and telephone calls via space. But Telstar had prematurely gone out of service in November that year, thought to be due largely to the deleterious effects of atmospheric nuclear weapons tests. A workaround brought the satellite temporarily back to life in January 1963 but its transistors were finally overwhelmed a month later, and it shut down. This had freed up time on the Holmdel antenna for some radio astronomy.

Penzias and Wilson were looking for microwave radiation that was thought to be emitted by the glowing cloud of gas surrounding the Milky Way. They began in late May 1964 by pointing the antenna at the sky in a direction at right angles to where they expected to find the radiation, intending to establish a baseline against which they could eventually measure their signal. To their great surprise, they found a substantial signal, a baseline so large it would actually overwhelm the signal they were hoping to study.

Radio engineers tend to characterize noise in terms of a 'noise temperature' or 'antenna temperature'. From their measurements, they estimated an excess antenna temperature of 3.5 K, plus or minus 1 K. It's worth noting that the 3.5 ± 1 K result was literally an 'excess' noise. Penzias and Wilson were *expecting* noise with a temperature of 2.3 K due to atmospheric absorption at zenith (with the antenna pointing directly upwards) plus another 0.9 K arising from absorption in the walls of the antenna. What they actually got was a noise temperature of 6.7 K. They were not the first to notice the discrepancy, and it had remained a mystery, a 'dirty little secret' among those Bell Laboratory radio engineers familiar with it.[27]

Photo 9 Robert Wilson (left) and Arno Penzias (right) were puzzled by noise picked up by the 20-foot microwave horn antenna at the Bell Telephone Laboratories' Holmdel research facility in New Jersey. It appeared to be coming from all directions in the sky.

For the kinds of wavelengths and intensities being studied, the antenna temperature is equivalent to the black-body temperature of the radiation.

Penzias and Wilson set about the task of identifying the source of this unwelcome noise so that they could eliminate it. The radiation did not seem to depend on direction—the signal was coming uniformly from all directions in the sky, with a uniform temperature. The signal didn't change with the time of day or season. They ruled out microwave pollution from nearby Manhattan. They evicted a pair of pigeons that were roosting inside the horn and cleared their droppings from its surface. But these were homing pigeons and, when they inevitably returned, the astronomers had to adopt a rather more permanent method of removal. Alas, the pigeons died in vain, as the source of microwaves persisted.

No nearer to solving their problem, in December 1964 Penzias attended a meeting of the American Astronomical Society in Montreal. He mentioned these puzzling observations to Burke, and in late February 1965 Burke advised him of an interesting talk by 'a guy from Princeton' in which he had predicted a cosmic black-body radiation temperature of 10 K. Burke sent Penzias a copy of Peebles' unpublished manuscript. Penzias immediately picked up the phone and called Dicke. The call 'lasted for some considerable time'. Penzias invited the Princeton physicists to take a first-hand look at the antenna on Crawford Hill, and to examine their data.[28]

Dicke, Roll, and Wilkinson piled into a car and drove the short distance to Holmdel. Penzias suggested publishing a joint paper on the discovery, but Dicke declined. They agreed to publish two short papers back-to-back in the letters section of the *Astrophysical Journal*. The paper by Dicke, Peebles, Roll, and Wilkinson titled 'Cosmic Black-body Radiation' was received by the journal on 7 May 1965.[29] It describes the theory and includes a Princeton version of a divine curve (see Fig. 21), calculated to reflect a cosmic background temperature of 3.5 K and an age of about 6 billion years. The paper references Alpher–Bethe–Gamow, and a 1953 paper by Alpher, Herman, and James Follin. But the 1949

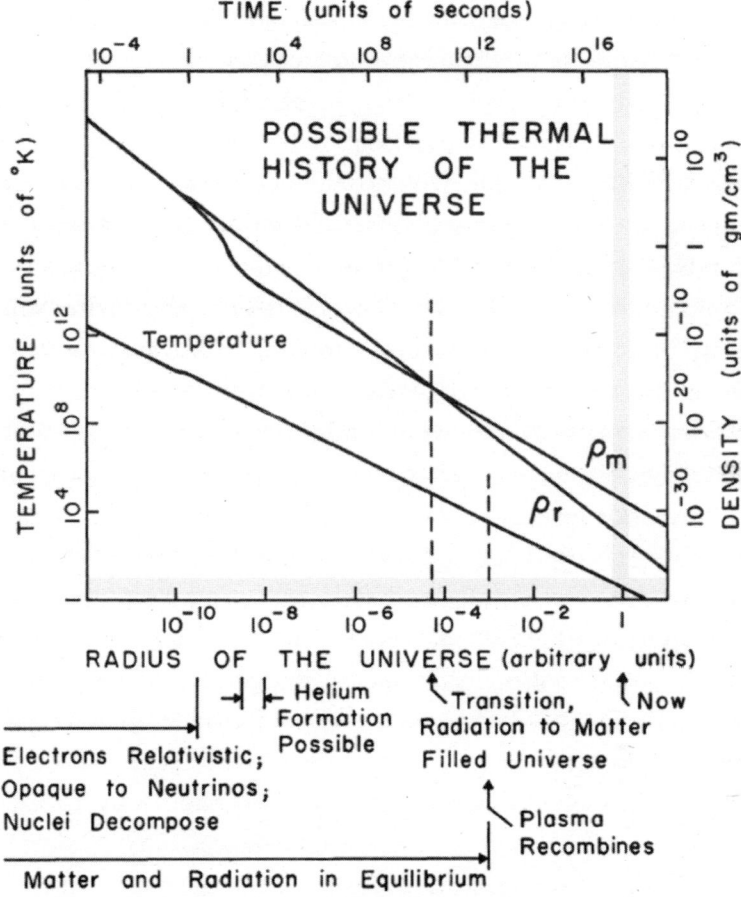

Fig. 21 This 'divine curve' was featured in the companion paper to the Penzias–Wilson announcement of excess antenna temperature at 4.08 GHz.

Alpher–Herman prediction of a temperature of 5 K was overlooked once again. Dicke took the blame, but Peebles later accepted his share. 'Our ignorance in 1964 about the literature of this subject is legendary in the cosmology community, and legends beguile ... I have the greater share of blame for poor homework: Bob [Dicke] was careful to stand back and let younger people in his group get on with research on their own'.[30]

As the title suggests, the paper by Penzias and Wilson, 'A Measurement of Excess Antenna Temperature at 4080 Mc/s', received on 13 May, stuck to the bare facts of the experiment.[31] It makes no attempt to interpret the signal, other than to acknowledge 'A possible explanation ... given by Dicke, Peebles, Roll, and Wilkinson' in a companion paper. Penzias sent a copy to Gamow, who replied tersely with a short summary of the predictions that he, Alpher, and Herman had made. 'Thus, you see the world did not start with almighty Dicke'.[32] It seems that Walter Sullivan may have had a mole in the editorial office of the *Astrophysical Journal*. Just eight days after the Penzias–Wilson paper was received, news of the discovery was splashed on the front page of the *New York Times*, with the headline 'Signals imply a "Big Bang" universe', and with a photograph of the 20-foot horn antenna at Holmdel.[33]

Everything changed. The steady-state model retained a few diehard defenders, some of whom argued that the background radiation must be due to scattered starlight. There was a lot more to be done to characterize the radiation and confirm its properties. But the oldest light, witness to events that had occurred when the universe was in its infancy, was getting ready to testify.

6

Hubble Wars

It was important to discover if the spectrum of the radiation detected by Penzias and Wilson was that of a genuine black body. This is produced only when the source and the radiation reach equilibrium. Such radiation is said to be 'thermalized'. Even the most rudimentary glance at the universe today—with its decidedly unequal sprinklings of matter and light in different directions across the sky—suggests that any radiation produced more recently in its history is unlikely to be thermal, at least in the context of the entire universe. Only the conditions prevailing in a hotter, denser, post-big bang universe can explain why we might expect a cosmic black-body spectrum. But 'spectrum' implies measurements over a range of wavelengths, and a single wavelength obviously doesn't qualify.

Fortunately, corroborating evidence was quickly forthcoming. The temperature of the cosmic background is essentially the temperature of 'space' throughout the universe. A number of scientists independently realized that the spectra of free cyanide (CN) radicals in interstellar clouds could be used to reveal their temperatures, which were expected to be the same as the background. Subsequent measurements all indicated a background temperature of about 3 K. And, although they had been scooped, the Princeton group continued with their own experiment and in 1966 reported measurements at a wavelength of 3.2 cm. Penzias and Wilson repeated their measurements at 7.35 cm using a different antenna, and then switched to 21 cm, a wavelength studied by another group at around the same time. All the results were consistent with cosmic black-body radiation with a temperature of 3 K.

But these were wavelengths far removed from the peak in the black-body spectrum, which for a temperature of 3 K is expected at around 1 mm (0.1 cm). To get closer to this peak, it was necessary to use a more sensitive radiometer and to get above much of the water vapour in the atmosphere by carrying it to the top of a mountain. The Princeton group, now consisting of Wilkinson, Bruce Partridge, Robert Boynton, and Paul Stokes, performed experiments at the White Mountain Research Station in the Sierra Nevada and at the High Altitude Observatory in the Rocky Mountains. They were able to gather data at four different wavelengths, from 3.2 cm to 0.33 cm (see Fig. 22). Although any conclusion drawn from the data depended on the result at the shorter wavelength (which produced the largest error bars), they were at least beginning to see some curvature in the pattern, consistent with a black-body spectrum with a temperature of 2.7 K.[1]

Any further refinements would require carrying the apparatus higher than the tops of mountains or—better still—above the atmosphere.

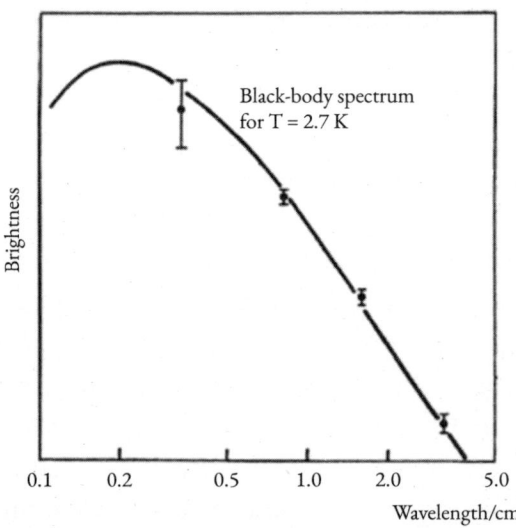

Fig. 22 In the late 1960s measurements of the cosmic background radiation performed at the tops of mountains in the Sierra Nevada and Rocky Mountains hinted at a black-body spectrum with a temperature of 2.7 K.

Steps to the Hubble Constant

The cosmic distance scale had been the third item listed on Hubble's cosmological programme for the Hale telescope. It was also crucial to reducing the uncertainty in the value of the Hubble constant. As before, the first rung of the distance ladder was based on the period–luminosity relation for the Cepheids that had been discovered by Leavitt. Baade's correction and further adjustments by Sandage had shown just how acutely sensitive this is, and it was clear that there was more work to be done. But the Cepheids are individual stars, and could be used as standard candles only within the Local Group of galaxies and within a neighbouring group of 34 galaxies in the constellation Ursa Major, called the 'M81 group'. This includes M81 itself (Bode's galaxy) and NGC 2403, and lies about 4 Mpc distant. At this distance, a Hubble constant of 75 km/s/Mpc implies a recession speed of just 300 km/s, which is about the same order of magnitude as the local (non-cosmological) 'peculiar' motions of galaxies (recall that the Andromeda galaxy is moving towards the Milky Way with a speed of 300 km/s). Consequently, knowing the distances of these local galaxies doesn't aid the accurate determination of the Hubble constant. The problem was that Cepheids lying beyond this distance must surely exist in distant galaxies, but were simply too faint to be observed.

Sandage judged that galaxies observed at distances beyond about 100 Mpc could be assumed to be moving uniformly with the 'Hubble flow', meaning that their redshifts and apparent recession speeds are overwhelmingly cosmological in origin. With recession speeds of 5000 km/s, any local motions of these galaxies would be swamped. At this distance, the brightness of entire galaxies was being measured, rather than individual stars. The challenge was to find a way of reliably constructing the next rungs of the ladder, connecting distances to the Local Group and the M81 group with the distances of galaxies in the Hubble flow. Once the Hubble constant had been firmly established, the Hubble–Lemaître law could then be used to determine the distance of a galaxy from its redshift.

But it was first necessary once more to address the period–luminosity relation of the Cepheids. Following Baade's correction and Sandage's adjustments, a number of attempts had been made to extend the relation to include the colours (and hence temperatures) of the variable stars, and from the late 1950s a series of period–luminosity–colour (PLC) relations were published. Although it was not immediately apparent, these relations were not so much about the intrinsic colours of the stars but were rather about the reddening effect of interstellar gas and dust along their lines of sight.

By the early 1960s, Sandage had collected hundreds of photographic plates of nearby galaxies taken over a 10-year period using the Hale telescope. Like Pickering many years before him, he needed to find someone to assist in their analysis, and settled on a young German astronomer called Gustav Tammann, grandson of a Russian physical chemist (also Gustav) who had fled Russia in 1903 and settled in Göttingen in Germany. After the death of his father, Tammann had moved with his mother to Basel in Switzerland, where he studied astronomy at the university. Sandage brought him to Pasadena in 1963, and their collaboration would span almost 50 years.

Identifying Cepheids from the plates that Sandage had accumulated was tedious work. But five years later Sandage and Tammann published a simple period–luminosity relation based on more than 100 Cepheids in the Large and Small Magellanic Clouds, the Andromeda galaxy, and NGC 6822 (Barnard's galaxy), all in the Local Group.[2] They used Cepheids within the Milky Way to fix the zero point. The data suggested that the relation remains consistent from galaxy to galaxy, in terms of both the slope and zero point, meaning that this is very much a characteristic of the Cepheids themselves. The results are shown in Fig. 23. Although this graph shows some slight curvature at periods of 100 days or more, the line is pretty straight otherwise and can be reduced to:[3]

$$\langle M \rangle = -1.43 - 2.80 \times \log P$$

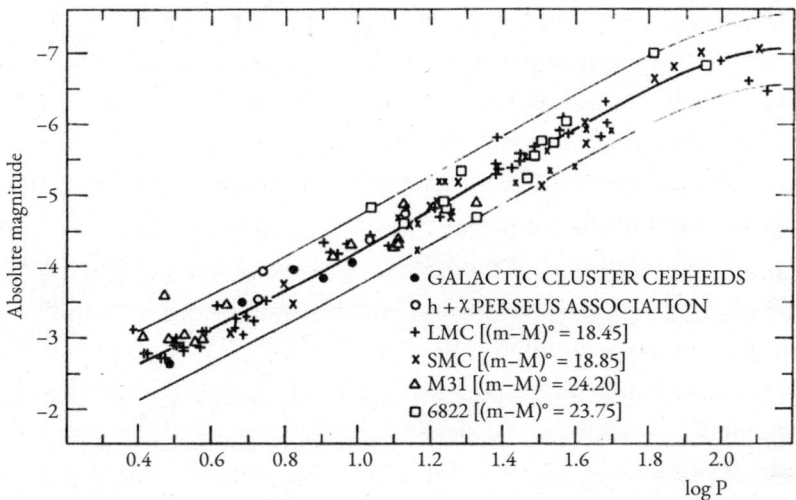

Fig. 23 Sandage and Tammann recalibrated the Cepheid period–luminosity relation in 1968. This put the distance modulus of the Andromeda galaxy (M31) at 24.20, corresponding to a distance of 691,800 parsecs or 2.256 million light-years (about 92% of the currently accepted distance).

which can be compared with Hertzsprung's original calibration from 1913. The Sandage–Tammann calibration put the distance to the Small Magellanic Cloud at about 59,000 parsecs (vs 10,000 parsecs based on Hertzsprung and 19,000 parsecs based on Shapley). Today this distance is determined to be 62,400 parsecs with an uncertainty of plus or minus 500 parsecs.

Sandage and Tammann decided that the features that had caused Hubble so much grief in the 1930s could provide the next rung on the distance ladder. These are the H II regions, whose diameters had been found to be correlated with the sizes of their host galaxies. The sequence of steps runs as follows. First, calibrate the diameters of the H II regions in selected galaxies using distances derived from the Cepheid variables. Second, extend this calibration to galaxies lying beyond 4 Mpc using a number of different methods, including individual distance indicators such as the magnitudes of the brightest blue stars.

Third, use the H II region diameters to determine the distances to 50 galaxies up to a maximum distance of 25 Mpc, and thence determine the statistical distribution of absolute magnitudes in this sample. In 1960, the Dutch-Canadian astronomer Sidney van den Bergh at the University of Toronto had derived a relation between the development of a galaxy's spiral arms and its luminosity, and had assigned galaxies classified as spiral and irregular to nine luminosity classes. For example, a spiral galaxy with the Hubble classification Sc ('late' spirals with a less pronounced central bulge and loosely wound arms) could now be further classified according to luminosity as Sc I, Sc II, etc. The third step then involves mapping the distribution of absolute magnitudes of the galaxies using this classification.

Fourth, use this distribution to calibrate the absolute magnitudes of newly identified giant Sc I galaxies up to a distance of 100 Mpc. This assumes that the 50 galaxies in the middle distance represent a fair sample of galaxies across the entire universe. The redshifts of these galaxies could then be used to make an initial determination of H_0. Sandage and Tammann published their results in a series of papers titled 'Steps towards the Hubble constant'. The third in the series, published in 1974, offered the value $H_0 = 56 \pm 9$ km/s/Mpc, a factor of 10 smaller than Hubble's estimate from 1929. Subsequent papers suggested 57 ± 6, and 57 ± 3 km/s/Mpc, giving a Hubble age of 17.5 billion years.

Open Hostilities

On the surface, this looked like a considerable improvement in accuracy and precision, but not all astronomers were convinced. Van den Bergh and French astronomer (and adopted Texan) Gérard de Vaucouleurs argued that the distance ladder that Sandage and Tammann had constructed appeared questionable in places. Loosen any rung even slightly and the whole thing was in danger of falling apart.

One of van den Bergh's graduate students, Barry Madore, argued that the faintness of NGC 2403 was not entirely due to its distance. The brighter, longer-period Cepheids were also younger, and so more likely to be shrouded in clouds of gas and dust. He put the distance of NGC 2403 at 2.2 Mpc, not 3.2 Mpc as Sandage and Tammann had determined in their first 'Steps' paper. Sandage rejected his arguments: 'He [Madore] came and spent time with me and I told him he was wrong and he went and published a devastating paper against Tammann and me'.[4] Madore was not completely wrong—the current accepted distance to NGC 2403 is 2.96 Mpc, but this was just the beginning of an assault on Sandage and Tammann's efforts that would quickly spiral into open hostilities.

Heated disputes are rare but not unusual in science, and often bring an element of psychodrama to scientific conferences that are more typically rather dull. They tend to leave a trail of seething anger and bruised and battered egos, and although scientific standards and the pursuit of truth are (sometimes) beneficiaries, such disputes are rarely resolved to the satisfaction of the protagonists, or interested onlookers.

De Vaucouleurs had carefully studied Sandage and Tammann's 'Steps' papers. He found their arguments confused and often circular, and identified what he claimed were 12 'blunders'. He argued that their programme was altogether too dependent on a few 'precision indicators', specifically the Cepheid period–luminosity relation, whose accuracy could be questioned. This was compounded through their use of a single calibration method. In contrast, de Vaucouleurs preferred a philosophy of 'spreading the risk', drawing on a multiplicity of independent methods.[5]

Furthermore, there was no good reason to suppose the universe is as uniform as Sandage and Tammann had assumed, with an even and equitable distribution of galaxies across the sky. Zwicky had argued as long ago as 1938 that galaxies (then called nebulae) tend to be found in clusters. In 1951 a young Vassar graduate called Vera Rubin had reanalysed the radial velocities of more than 100 galaxies as part of her Master's degree at Cornell University, attempting to unpick from them a rotational motion of the entire universe. She concluded that galaxies were indeed moving at speeds which diverged from the Hubble–Lemaître law, some much faster, some much slower.[6]

The Palomar Observatory Sky Survey, funded by the National Geographic Society, began collecting the first of its 2000 photographic plates in November 1949, and concluded in December 1958. One of the study's principal observers, George Abell, produced a catalogue of 2712 'rich' galaxy clusters, in which he divided the clusters into a series of 'cluster groups', ranging from 30–49 galaxies to over 299 galaxies. The catalogue formed part of his Caltech PhD thesis, in which he concluded: 'The data suggest strongly the existence of second-order clusters, or clusters of clusters of galaxies'.[7] A further 1361 rich clusters were contributed in the 1970s by the Southern Sky Survey, conducted using the UK Schmidt telescope at the Anglo-Australian Observatory in New South Wales.

Although Rubin's efforts were poorly received at the time, de Vaucouleurs confessed himself to be 'more pugnacious'.[8] He drew courage from Rubin's work and was familiar with Abell's thesis, and in 1958 went on to argue that the Local Group is merely an outlier of a local *supercluster* of galaxies with a diameter between 20 and 30 Mpc.[9] This is now known as the Virgo supercluster, which consists of at least 100 galaxy clusters. He would later emphasize that a supercluster is not the same as Abell's 'cluster of clusters': it is much larger. In 2014, astronomers concluded that the Virgo supercluster forms a lobe of an even greater supercluster named Laniakea (Hawai'ian for 'open skies' or 'immense heaven').

The universe is not uniform: it is rather characterized by matter drawn into long filaments and walls, surrounded by deep voids.

De Vaucouleurs chose to launch his offensive at a high-profile presentation to the 16th General Assembly of the IAU, held in Grenoble, France, in late August and early September 1976.[10] Sandage and

Tammann were wrong: the value of the Hubble constant was more like 100 km/s/Mpc, the universe was half the size they had claimed, and half the age. He went on to develop a rival system, not so much a distance ladder but a substantially more complex structure drawn to resemble the Eiffel Tower.

De Vaucouleurs' views about the large-scale structure of the universe were judged to be little short of heretical, and the debate remained open.

Going Big

As the nursery rhyme suggests, little stars twinkle. They do this because the Earth's atmosphere can be turbulent, giving rise to what astronomers call 'bad seeing'. Although it wouldn't be completed for another three years, Princeton theoretical physicist and astronomer Lyman Spitzer had already concluded in 1946 that the 200-inch Hale telescope would reach the resolution limit imposed by such turbulence, and that larger telescopes would offer no appreciable advantage. Ground-based telescopes are also blind to infrared and ultraviolet light, which are absorbed by the atmosphere.

Although regarded by many as science fiction, the development of rocket technology during the second world war had opened up possibilities for Earth-orbiting satellites and a space-based observatory. Spitzer summarized the scientific advantages of a space telescope in a report prepared in July 1946 for the Douglas Aircraft Company's Project RAND, an organization formed after the war to connect research and development with military planning, based in Santa Monica, California. RAND is a contraction of Research ANd Development and, in 1948, this was established as an independent non-profit corporation. It continues operations today.

Spitzer acknowledged that 'a large reflecting satellite telescope (possibly 200 to 600 inches in diameter) is some years in the future', but went on to speculate on its advantages for astronomy and cosmology.[11]

He did not speculate on what it might cost, or how it might be built. As the report was classified, it remained unknown to the astronomy community for some years.

But the business of Project RAND was to *anticipate* the future. In a follow-up to its first report, issued in May 1946 and titled 'Preliminary Design of an Experimental World-circling Spaceship', the Head of Project RAND's Missile Division made this prescient observation: 'To visualize the impact on the world, one can imagine the consternation and admiration that would be felt here if the United States were to discover suddenly that some other nation had already put up a successful satellite'.[12] The launch of *Sputnik 1* by the Soviet Union in October 1957 (and *Sputnik 2* a month later) shattered the American public's perception of US post-war technological supremacy. The US responded with *Explorer 1* in January 1958, and the National Aeronautics and Space Administration was established the following July. It began operations in October. The 'space race' began.

In 1965 Spitzer was selected to head a NASA committee to determine the scientific objectives of what had been named as the 'large space telescope'. Although astronomers had been initially sceptical of the value of a space-based telescope, a National Academy of Sciences report in 1969 extolled its virtues for all types of astronomy, including studies of the Hubble constant and deceleration parameter. The first working group meeting of astrophysicists and engineers took place in 1974, and a budget was approved by Congress three years later.

Also in 1974, NASA issued two 'Announcements of Opportunity', inviting scientists to submit proposals for astronomical missions using small and medium-class *Explorer* satellites. Of the 121 proposals that were received, just three were concerned with the study of the cosmic background radiation. These were initially rejected in favour of the Infrared Astronomical Satellite (IRAS), but they were nevertheless judged by NASA to be worthy of further consideration. In late 1976 NASA requested that the three teams that had submitted proposals join forces and work together on a joint mission. A small Mission Definition

Study Team was formed, consisting of Samuel Gulkis at the Jet Propulsion Laboratory in Pasadena, Michael Hauser and John Mather at the Goddard Space Flight Center in Maryland, George Smoot at the University of California, Berkeley, Rainer Weiss at MIT, and David Wilkinson at Princeton.

It was agreed that the satellite would carry three experiments: a microwave radiometer to map the temperature of the cosmic background radiation across the sky (led by Smoot), a far-infrared absolute spectrophotometer to measure its spectrum (led by Mather), and a diffuse infrared background experiment which would look for relic radiation from the first stars and galaxies (led by Hauser). They chose to call the mission the Cosmic Background Explorer, abbreviated as COBE. In 1977, Steven Weinberg was making some final corrections to the proofs of a new popular book, *The First Three Minutes* when he received from Mather a copy of the first *Cosmic Background Explorer Satellite Newsletter*. He wished them *bon voyage*.[13]

There was much to be done, and it would take many years, but the science of cosmology was getting ready to go big.

A Goldilocks Universe?

For the seventies-era cosmologist, the presumed uniformity of the universe was both a blessing and a curse. Einstein had founded the new discipline of relativistic cosmology in 1917, simplifying the mathematical problem by adopting the cosmological principle: the universe is homogeneous and isotropic in all directions. Uniformity was a boon: it had allowed cosmologists to calculate their 'divine curves', and Sandage and Tammann had assumed a high degree of uniformity to assemble their steps towards the Hubble constant. Isotropy appeared to be literally baked in to the cosmic background radiation, which to all intents and purposes looked like the thermal radiation of a black body, uniform across the sky.

So, what's the problem? If we look up at the night sky, we can perhaps accept that, in a very coarse-grained sense, the universe does indeed look much the same in different directions. We see stars, faint wisps that we call nebulae, and the blackness of empty space. However, a child, unfazed by any scientific authority, will simply point out that the pattern of points of light over here on the left looks very different compared with the pattern of points of light over there on the right. In a fine-grained sense, we have to own up to the fact that the universe has a *structure*. This goes far beyond what we can see with the naked eye. If de Vaucouleurs was right, this is a structure that stretches over millions of parsecs, involving millions or billions of galaxies, a structure that is perhaps rather incompatible with an isotropic universe.

The key problem could be reduced to a simple question, one that cosmologists were beginning to realize demanded an answer from any hot Big Bang cosmology worthy of the name. *Why do stars and galaxies exist?*

In 1969 Dicke had sought to draw attention to the extraordinary sensitivity of Big Bang cosmology to its initial conditions. 'If the [Big Bang] fireball had expanded only 0.1 per cent faster, the present rate of expansion would have been 3×10^3 times as great. Had the initial expansion rate been 0.1 per cent less, and the Universe would have expanded to only 3×10^{-6} of its present radius before collapsing. At this maximum radius the density of ordinary matter would have been ... over 10^{16} times as great as the present mass density. No stars could have formed in such a Universe, for it would not have existed long enough to form stars'.[14]

Ten years later, Dicke and Peebles deepened the mystery. They asked what the total matter density would need to be just one second after the Big Bang, when primordial nucleosynthesis was thought to have begun, in order to produce a spatially flat, Einstein–de Sitter universe. They concluded that the matter density must have been equal to the critical density ρ_c to within an astonishing one part in 10^{14}. A density just a factor of one hundred trillionth larger than ρ_c would have resulted in a closed universe which would have collapsed long before stars and galaxies could form. A density just a factor of one hundred trillionth smaller than ρ_c would have resulted in an open universe with no opportunity for stars and galaxies to form. Of course, the matter density was whatever it was one second after the Big Bang and the cosmological models could not predict this from first principles. But they might at the very least be expected to provide some kind of answer to the question: why *this* density and no other? Dicke and Peebles referred to this as the *flatness problem*.[15]

This was somewhat exacerbated by the fact that when cosmologists looked for all this matter, what they found in terms of visible stars and galaxies fell far short of the critical density. In 1958, Oort had estimated a matter density of 7×10^{-31} g/cm³ (about one hydrogen atom on average in every couple of cubic metres of space). The critical density ρ_c can be calculated from the Hubble constant.[16] Assuming $H_0 = 50$ km/s/Mpc (Sandage and Tamman) gives $\rho_c = 4.7 \times 10^{-30}$ g/cm³, such that $\Omega_M = \rho/\rho_c = 0.15$. Assuming $H_0 = 100$ km/s/Mpc (de Vaucouleurs, and others) gives $\rho_c = 1.9 \times 10^{-29}$ g/cm³, and $\Omega_M = 0.04$. Both are far short of the magic ratio $\Omega = \Omega_M = 1$.

Flatness is a large-scale problem, but there was more trouble at smaller scales. If, however it came about, the critical density of matter is distributed uniformly in an expanding space, gravity is expected to pull on it equally from all directions, such that nothing will happen—see Fig. 24(a). What the early universe needed was a little inhomogeneity or *anisotropy*. A small anisotropy, leading to an excess of matter in a region of space (called an over-density), may be sufficient to allow gravity to overcome the effect of expansion, drawing matter together. Gravity is irresistible, and over time the effect is amplified—Fig. 24(b).

We tend to think of expansion as a property of the entire universe, but we can also think of it as a local phenomenon. We can apply the same logic to a spherical volume of space *within* the universe. If the density of matter in such a sphere exceeds the critical density, the sphere is 'closed'—over time it will collapse in on itself, providing opportunities to form stars and galaxies along the way. If instead the mass inside the sphere has a density below the critical density, then it is 'open' and expansion will continue unchecked. No stars or galaxies can form and the result is a void. So, even if the early universe contained exactly the critical density of

(a)

(b)

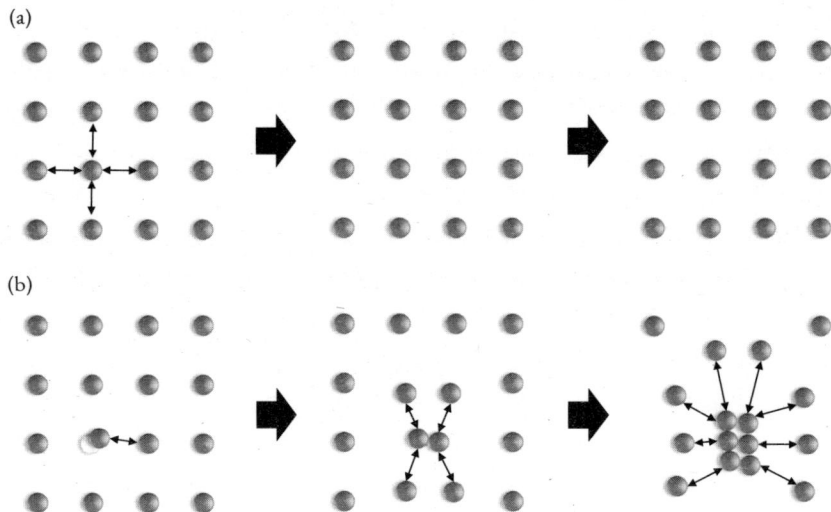

Fig. 24 (a) In a perfectly uniform or isotropic distribution of matter gravity acts equally in all directions and the matter remains frozen in place, even though space itself is expanding. In (b) a small anisotropy in the distribution of matter (an over-density) creates a slightly stronger local gravitational attraction. Over time the matter clumps together, forming a structure.

matter, this had to have been distributed unevenly. We can think of these over-densities as 'cosmic seeds' that will lead eventually to the formation of stars and galaxies.

But there were several problems with this. The most conspicuous was that, like the flatness problem, it requires what seems to be an extraordinary fine-tuning. Make the local over-densities too large and all the matter in the universe would have long ago collapsed to an ultra-dense state. Too little density would mean no stars or galaxies at all. What was needed was just the right sprinkling of over-densities of just the right size: a 'Goldilocks' universe.

The theoretical study of anisotropy in the distribution of matter in the early universe was begun in 1946 by Russian physicist Evgeny Lifshitz.[17] In 1970, British cosmologist Edward Harrison at the University of Massachusetts, Amherst and, independently in 1972, Russian Yakov Zeldovich at the Institute of Applied Mathematics in Moscow, identified the scale of the density fluctuations necessary to produce the kind of large-scale structure of the universe we see today. The 'Harrison–Zeldovich spectrum' is characterized by density fluctuations of the order of 10^{-4} (0.01% or 1 part in 10,000).[18] These are 'scale invariant'—the extent of the fluctuations are the same irrespective of the scale (such as length) used to study them.

We can get some sense of this by thinking about fractals such as the Mandelbrot set. As we zoom in, we see the same patterns reproduced at smaller and smaller scales (but note—fractals are 'self-similar' rather than strictly scale invariant). Fluctuations of just 0.01% might seem small, but this is more than enough anisotropy for gravity to go to work on. But, why 0.01% and not some other figure?

To make matters worse, there was a further dimension to the question of why specifically spiral galaxies exist. As part of his post-doctoral research programme at Cambridge University in England, American physicist Jeremiah Ostriker had developed computer simulation techniques to study the stability of flattened, rotating stars. In 1965 he was appointed by Spitzer to a research associateship at Princeton Observatory, and became a full professor at Princeton University in 1971. Wondering if the same physical principles might apply to flat, rotating spiral galaxies, he stopped by Peebles' office to ask what he thought. They collaborated on a simulation model, and concluded that the principles were indeed the same.

They also concluded that, based on the mass of its visible stars, a spiral galaxy like the Milky Way is inherently (and dramatically) unstable. The simulations suggested that it should either collapse into a bar-like structure or split into a binary system. As far as the physics was concerned, the Milky Way, and all other spiral galaxies, should not exist.

Dunkle Materie

In their paper, published in 1973, Ostriker and Peebles offered an explanation of why spiral galaxies do indeed exist. They concluded that there must be more to a spiral galaxy than meets the eye—a flattened, rotating galaxy could be stabilized if it were embedded in a large spherical 'halo' of matter that, for some reason, is invisible.[19] In a further paper co-authored a year later with Amos Yahil, Ostriker and Peebles suggested that the masses of spiral galaxies deduced from their visible stars underestimated their actual masses by a factor of 10 or more.[20]

Ostriker discovered that this was not a novel deduction. Following a lecture delivered to the US National Academy of Sciences in 1976, he was approached by an elderly gentleman carrying a thick black book. His name was Horace Babcock and the book was his doctoral thesis, written in 1937, a year before Ostriker was born.[21] Babcock had studied the rotational motions of stars in the Andromeda galaxy. The results were puzzling. The galaxy's outermost stars were rotating around the centre of the galaxy at speeds much higher than predicted based on the mass of the galaxy that could be deduced from all its visible stars.

The logic is fairly simple. The shapes of spiral galaxies like Andromeda betray the fact that they are rotating, trailing swirls of stars, dust, and gas from their spiral arms. The population of stars in such a galaxy is most dense at the centre, suggesting that the gravitational field is strongest here. Stars close to the centre are expected to rotate much faster than stars lying further out, where the gravitational field is weaker.

We can figure out what this means by looking at our own solar system. Most of the mass of the solar system is concentrated at the centre (in the Sun) around which orbit the planets. Planets close to the centre feel the full force of the Sun's gravity and are whirled around at high speeds. Mercury, with an average orbital radius of about 58 million km, has an orbital speed of nearly 50 km/s. But as we get further and further away from the centre, the effects of the Sun's gravitational field weaken, and so we would expect more distant planets to orbit more slowly. Neptune, with an average orbital radius of about 4500 million km, has an orbital speed of only 5.4 km/s.

A spiral galaxy is a little more complicated. There are lots and lots of stars orbiting at considerable distances from the centre, each contributing to the galaxy's gravitational field. Instead of peaking at a fairly sharp point right at the centre and falling away, the field is more like a bell curve: it falls away gently at first, more strongly as the distance increases. We should therefore expect that the *rotation curve*—the graph of orbital speed vs radius—should rise to a peak not far from the centre before falling away. But this was not what Babcock had found.

Ostriker had been unaware of Babcock's work, which was published in the *Lick Observatory Bulletin* in 1939.[22] Babcock's thesis adviser was Nicholas Mayall and, with American astronomer Lawrence Aller, Mayall had extended the study to include the rotation curve of the Triangulum galaxy (M33), part of the Local Group. With Arthur Wyse, an assistant astronomer at Lick,[*] Mayall analysed the distribution of mass in Andromeda and Triangulum as revealed by their rotation curves. They observed that for Andromeda, the ratio of mass-to-light varies significantly with distance from the centre: and is 'unexpectedly large for the outer parts of the spiral. This result naturally raises the important question as to what kind of faintly luminous matter may be present in order to produce the large mass indicated by the observations'.[23] There appeared to be far more matter at the outskirts of the galaxy than was revealed by the brightness of its stars.

The irascible Zwicky hadn't thought that the matter was 'faintly luminous'. He called it *dunkle Materie*—dark matter. In his 1933 analysis of the mass of the nearby Coma cluster, a large galactic cluster located in the constellation Coma Berenices which contains over 1000 galaxies, he had deduced that the cluster must consist of about 400 times more matter than its total luminosity suggested. He wrote: 'If this were to prove to be true, the surprising result would be that dark matter is present in much greater density than luminous matter'.[24] Zwicky had studied a cluster of galaxies rather than a single spiral galaxy, but the principles are the same.

Though Zwicky is rightly credited as the discoverer of dark matter, by the mid-1970s his work had been all but forgotten.

---

[*] Wyse was among 12 crew and scientists killed in June 1942 in an accidental collision of light aircraft off the New Jersey coast during a wartime experiment.

It all seemed faintly ridiculous. Matter is constructed from positively charged protons and neutral neutrons (collectively called *baryons*) and negatively charged electrons. By its very nature, the stuff that results—predominantly hydrogen atoms and some helium atoms—interacts with light. The atoms absorb light and they emit light. Whatever it is, dark matter doesn't do this. It is characterized only by its gravity, and is otherwise utterly inscrutable. High-energy particle physicists had also been busy in the 1970s developing what would become known as the 'standard model' of particle physics. Ideas would later abound, but there was nothing in the standard model that could explain dark matter.

The older generation of astronomers like Sandage grumbled about it, but the evidence would soon become quite overwhelming. After completing her Master's degree, Rubin had moved to Georgetown University where she studied for a PhD under Gamow's supervision. She understood what she was getting herself into. Gamow simply threw out interesting questions and left his students to figure out how to try to answer them. Her thesis drew her back into controversial territory—galaxies were clumped together rather than uniformly distributed, as de Vaucouleurs continued to insist. Installed at the Carnegie Institution's Department of Terrestrial Magnetism, by the early 1970s she was ready for a research project that was more conventional and uncontroversial. 'I decided to pick a problem that I could go observing and make headway on, hopefully a problem that people would be interested in, but not so interested [in] that anyone would bother me before I was done'. She had by this time gained an important collaborator, Kent Ford, who had designed and built an image tube spectrograph 'that could do things that no other spectrographs could do'.[25] She chose to study the rotation curve of the Andromeda galaxy.

By 1978, Rubin, Ford, and their colleagues had measured rotation curves for 10 high-luminosity spiral galaxies. By 1980, they had measured rotation curves for 21 'late'-type spiral galaxies. All showed a similar pattern of behaviour: the rotation speeds increase to a maximum before levelling off or continuing to *increase* slightly (Fig. 25). 'The

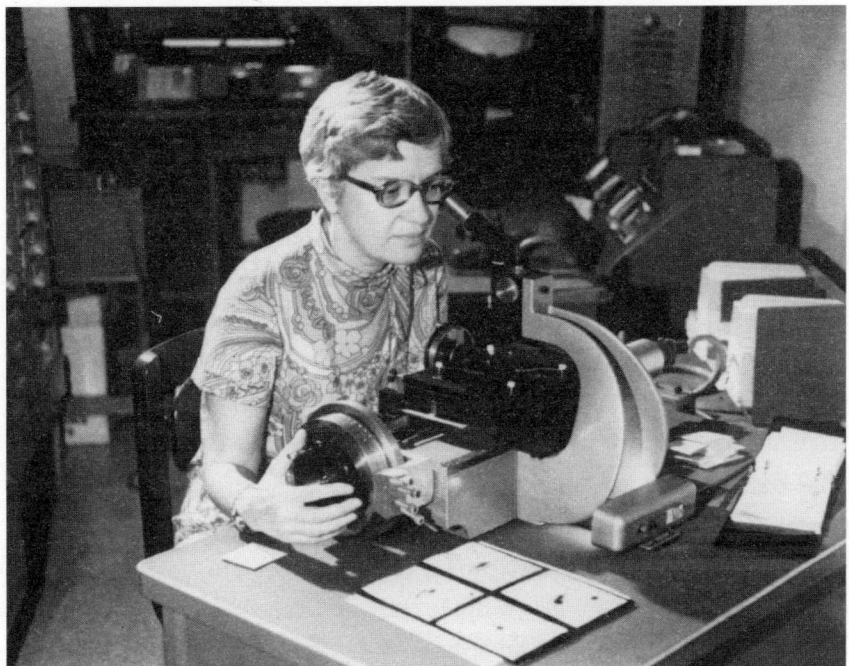

Photo 10 Vera Rubin measuring spectra at the Carnegie Institution's Department of Terrestrial Magnetism.

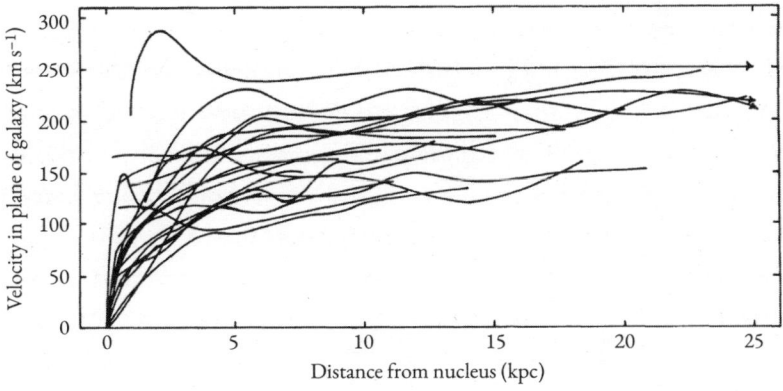

Fig. 25 Rotation curves for a series of 21 Sc galaxies reported by Rubin, Ford, and Thonnard in 1980. Based on the estimated mass of visible stars, these curves should peak a short distance from the galactic nucleus before declining. The observed behaviour is evidence for the existence of dark matter.

conclusion is inescapable that non-luminous matter exists beyond the optical [visible] galaxy'.[26]

The status of dark matter evolved quickly. It had started as 'missing mass' in a large galactic cluster, and was rediscovered as 'faintly luminous' matter missing from the Andromeda and Triangulum galaxies. It had then been rediscovered again as an essential ingredient required to stabilize spiral galaxies. In 1978 British astrophysicists Martin Rees and Simon White at Cambridge University suggested that dark matter is actually essential to galaxy *formation*.[27] Given its dominance, the cosmic seeds must have largely been over-densities in the distribution of dark matter, which was then drawn together by gravity to form haloes. The heavier baryonic matter that was sprinkled through the universe would have become concentrated at the centres of these haloes, eventually accumulating sufficient mass to form stars and galaxies. Part of the answer to the question 'Why do stars and galaxies exist?' could now be tentatively answered, but only at the cost of another unanswered question: What is dark matter?

Alas, dark matter could not resolve the problem with the density parameter. Ostriker, Peebles, and Yahil estimated that Ω_M, now understood as the sum of the density parameters of baryonic matter (Ω_B) and of dark matter (Ω_{DM}), was of the order 0.2, with an uncertainty not less than a factor of 3. In their calculations, White and Rees had assumed a model with $\Omega_M = 0.2$, with dark matter accounting for 80% of the total. This was still far short of the value $\Omega_M = \Omega_B + \Omega_{DM} = 1$ required for a flat universe. Dark matter could not solve the flatness problem.

The 'Fishy–Tuller' Relation

In July 1975, Canadian-born astronomer R. Brent Tully and American J. Richard Fisher submitted a paper to the journal *Astronomy and Astrophysics*. As graduate students at the University of Maryland, they had some years earlier developed a new method to determine the luminosities, and hence distances, of distant spiral galaxies. Relatively simple Newtonian physics suggests that the greater the mass of a galaxy, the faster it will rotate. And the greater the mass, the greater the number of stars and hence luminosity. When calibrated by reference to nearby galaxies, they surmised that it should be possible to relate the luminosity directly to the rotation speed. Measure the rotation speed of a more distant galaxy, and the *Tully–Fisher relation* could be used to infer its luminosity, or absolute magnitude.

This is an empirical relation which does not require the determination of the mass of the galaxy. Tully and Fisher did not include any consideration of dark matter in their analysis—this was a concept still very much under debate. But, as it happens, their method is aided by the presence of the dark matter halo which, as we've seen, tends to flatten the rotation curve. Measurement of a single 'rotation speed' for a spiral galaxy then corresponds broadly with the plateau in the rotation curve. And, as the amount of luminous baryonic matter and dark matter are likely to be connected—the larger the dark matter halo the larger the galaxy's baryonic matter content—the relation is preserved. In fact, the Tully–Fisher relation is a little 'tighter' if luminosity is replaced by baryonic mass.

Tully and Fisher proposed to measure the rotation speed from the width of a single line in the emission spectrum of neutral hydrogen atoms, with a wavelength of 21 cm. The logic is again quite simple and by now familiar, as it relies on the Doppler effect. Emission from the side of the galaxy rotating towards us is blueshifted by an amount that depends on its speed. Emission from the side of the galaxy moving away is redshifted by the same amount. These different sides are not distinguished in the measurement and so the end result is a stretching or 'Doppler broadening' of the 21 cm line. The extent of the broadening is then an empirical measure of the rotation speed. Because not all galaxies can be viewed 'edge-on', a correction must be applied for those galaxies viewed at an angle.

They used galaxies in the Local Group, the M81 group, and the M101 group (the Pinwheel galaxy and its companions) to calibrate their relation and used this to derive distances to galaxies in the Virgo and Ursa Major clusters. They obtained a distance modulus for the Virgo cluster of 30.6 (a distance of 13.2 Mpc), and for the Ursa Major cluster of 30.5 (12.6 Mpc). From these they deduced a preliminary estimate for the Hubble constant of 80 km/s/Mpc.

In the fourth of their 'Steps' papers, Sandage and Tammann had determined a distance modulus for the Virgo cluster of 31.45 (19.5 Mpc). In their paper, Tully and Fisher had volunteered suggestions for the discrepancy, but there was no getting away from the fact that this was a direct challenge to the Sandage–Tammann distance ladder.

The Tully–Fisher paper was sent for peer review. The refereeing process for scientific papers is anonymous, but there was clearly something odd going on. Unusually, the paper was sent twice to the referees, and a revised manuscript was received by the journal in April 1976. By the time the paper was published in 1977, two years after its original submission, Sandage and Tammann had already published a paper discrediting Tully and Fisher's approach. 'Some people say there was skullduggery', Tully said.[28]

Sandage didn't have much faith in the Tully–Fisher relation, and would sometimes refer to it as 'Fishy–Tuller'. Sandage and Tammann reanalysed the data that Tully and Fisher had used and came up with their own version of the relation. Unsurprisingly, perhaps, they now found a distance modulus for the Virgo cluster using the revised relation of 31.4 (19.1 Mpc), consistent with their previous estimate, and a Hubble constant of 50.3 ± 4.3 km/s/Mpc. 'The agreement with the 21 cm method again provides a satisfactory test of the current distance scale', they concluded.[29]

But Tully and Fisher did not agree. Having caught sight of a draft of the Sandage–Tammann paper, they added a note to the proofs of their own. 'For the moment it can only be concluded that the extragalactic distance scale is in doubt'.[30]

The American astronomer John Huchra characterized this period as follows: 'By the late 1970s, this bimodality remained in the estimates of

H_0 and the middle ground was littered with the bruised and battered remains of young astronomers attempting to resolve the dispute between the two sides'.[31]

The 'Hubble Wars' dominated distance scale discussions for more than 20 years. It was resolved not just by new, more accurate and precise measurements, but by irresistible changes to the way modern astronomy is done. Sandage and de Vaucouleurs belonged to the older generation of astronomers who worked largely in isolation, each taking individual responsibility for devising their own observational and analytical methodologies. This inevitably left the astronomer vulnerable, exposed to the charge of selection bias in the event of disagreements which could quickly become personalized. Later generations involved in 'big' astronomy would of necessity be much more collaborative and open, sharing their data with the wider community as projects progressed, lessening (though not eliminating) the risk of personalized disputes.

It would transpire that neither Sandage nor de Vaucouleurs was right.

7

The Inflationary Universe

The various divine curves that cosmologists had devised suggested that as the universe expanded, its temperature fell. At a temperature of about 4000 K, any protons and electrons that come together to form neutral hydrogen atoms are almost immediately broken apart again under bombardment by energetic photons. Under these conditions the early universe is a plasma of charged particles and photons. The photons are scattered predominantly from the electrons in a process called Thomson scattering, named for the English physicist J.J. Thomson. This is a form of 'elastic' scattering, meaning that the kinetic energy of the charged particle and the frequency of the photon are not appreciably changed in the collisions.

It's as though the universe is shrouded in fog. But at temperatures around 3000 K, the photons no longer have sufficient energy to strip the electrons from neutral hydrogen atoms, and the atoms begin to accumulate. Cosmologists call this 'recombination'. The photons no longer interact with matter and so have nowhere else to go. They are 'released' to form the cosmic background radiation. The universe becomes transparent.

Photons released from this 'last scattering surface' appear to be thermalized, with a black-body temperature of 3000 K. Some of this light is visible, though of course there is nobody around to see it. Further expansion cools this radiation to its present-day temperature of 2.7 K. As the temperature of the radiation scales with cosmological redshift, we can calculate the redshift z as $T/T_0 - 1$, where T is the temperature at some time in the past, and T_0 is the temperature today. Setting $T = 3000$ K gives $z = 1110$, representing the oldest light that we can see directly. If we assume $\Omega = \Omega_M = 0.2$, and $H_0 = 50$ km/s/Mpc, then the 'look-back' time corresponding to this redshift is about 470,000 years after the Big Bang. A Hubble constant $H_0 = 100$ km/s/Mpc gives a slightly earlier look-back time of 320,000 years.[1] Any light emitted earlier than this is lost in the fog.

As the universe expands and cools towards the critical temperature of 3000 K, the protons and electrons are pulled together by their electrostatic attraction and the electrons act like a kind of glue, binding the photons and the positively charged matter together. Such is the proximity of the protons, electrons, and photons that they can be treated together as a single radiation-dominated 'fluid'. As the gravitational pull from each over-density in the distribution of matter builds (look back to Fig. 24(b)), the radiation struggles to escape and suffers both redshift and time dilation effects, collectively referred to as the *Sachs–Wolfe effect*, named for Rainer Sachs and Arthur Wolfe at the University of Texas in Austin. Recombination proceeds more slowly in regions of high matter density, and the radiation released therefore appears slightly hotter there. In fact, it is estimated to take more than 100,000 years for recombination to complete.

It was quickly understood that, however they had arisen, the cosmic seeds required for galaxy formation would be expected to be translated into small differences in the temperature of the background radiation across the sky. The key question was: How small? In 1967, Sachs and Wolfe predicted positive and negative fluctuations from the mean background temperature as high as 1%.[2] In the same year British astrophysicist Joe Silk at the Harvard College Observatory estimated the fluctuations at 0.03%.[3] Recall that the Harrison–Zeldovich spectrum suggested that to be consistent with the present-day universe, the density fluctuations needed to be of the order of 0.01%, implying similar fluctuations in the temperature of the cosmic background. The theorists needed an explanation for why the fluctuations should be so *small*. And, of course, for the model to hold together with our experience of the universe, it was necessary to *observe* them in the cosmic background.

Such small fluctuations would challenge the limits of sensitivity of a microwave radiometer. To complicate matters somewhat, there was another source of anisotropy expected at this level. This has nothing to do with any intrinsic anisotropy of the cosmic background radiation, but results from the simple fact that the Earth is moving through it, and is the result of a Doppler effect. The radiation is expected to be slightly hotter in the direction the Earth is moving through the background, and slightly cooler in the opposite direction, called the *dipole anisotropy*. The reason for this terminology will be revealed shortly.

Any attempt to discover evidence for cosmic seeds by mapping fluctuations in the temperature of the cosmic background radiation across the sky would first need to measure and subtract out the dipole anisotropy. Peebles and British physicist Dennis Sciama had estimated that if the Sun and Earth orbit the Milky Way galaxy with a rotation speed of 250 km/s, this should give rise to a dipole anisotropy of about 0.07%, or ± 0.002 K.[4]

The Dark Lady and the Dipole Anisotropy

George Smoot was in high school when the Soviet Union launched *Sputnik*. Jolted from its complacency, an embarrassed US administration passed the National Defense Education Act in 1958, committing a billion dollars to fund a new science curriculum. Smoot's mother provided him with additional tutoring on science and history, and his father force-fed him a diet of trigonometry and introductory calculus.[5] Although he dallied briefly with medicine, his path to physics was set. A degree in mathematics and physics and a PhD in particle physics at MIT followed.

In 1970, he was drawn to work with Luis Alvarez in Berkeley on a NASA-funded High Altitude Particle Physics Experiment (HAPPE, or 'happy'). Alvarez had chosen high-altitude experiments in the 1960s as insurance against a decision by Congress to cease funding larger particle accelerators. But this turned out to be unnecessary when Congress agreed to continue funding the National Accelerator Laboratory, which eventually became Fermilab. Despite its name, the excursion into high-altitude experiments did not end happily. At the end of its maiden flight, HAPPE lost its payload to the Pacific Ocean. Some rethinking was needed.

Smoot's arrival coincided with a shift in focus, away from particle physics to high-altitude studies of cosmic rays. The switch was not immediately auspicious. A search for an antimatter component in the cosmic rays found no evidence for it. In May 1977, the dramatic, high-velocity return of a high-altitude balloon to the Badlands of South Dakota signaled an abrupt end to the project. Smoot's team was able to recover data from the wreckage, but for Smoot himself: '[m]y infatuation with ballooning was over'.[6]

Fortunately, Smoot had earlier anticipated the need to find a new research project. One source of inspiration was a talk he had attended as a first-year graduate student at MIT given by Silk on his Harvard thesis topic (and the basis for his 1967 paper) in which he had posed the question: 'would fluctuations in the primordial fireball survive to an epoch where galaxy formation is possible?' As he later acknowledged: 'Silk's work also made me realize the enormously important role of the cosmic background radiation in the early universe'.[7] Silk had moved to Berkeley in 1970. A second source of inspiration was a section in Peebles' book *Physical Cosmology*, first published in 1971. As a first step, Smoot's research team elected to fix on a search for the dipole anisotropy.

Such a search would require unprecedented detection sensitivity, requiring a sophisticated descendent of the Dicke radiometer called a differential microwave radiometer (DMR). As the name implies, a DMR has two horn antennas and measures the *difference* in the temperature of the radiation in different directions. This would subsequently become one of the instruments considered for the COBE satellite mission, which lay some years in the future. To get as far up above the atmosphere as possible without going into orbit, Smoot considered the White Mountain Research Station and more high-altitude balloon experiments. But fortuitous circumstances led him to consider instead an airborne observatory installed on a C-141 military transport plane.

It quickly transpired that the C-141 was unsuitable, so Smoot settled on a Lockheed U-2 reconnaissance aircraft, capable of flying at altitudes of 70,000 feet, also known as the U-2 'spy plane'. This aircraft was operated by the US Central Intelligence Agency during the cold war and features in many cold war 'incidents'. CIA pilot Gary Powers was flying a U-2 over Soviet territory when he was shot down in 1960. Major Rudolf Anderson, Jr was flying a U-2 when he was shot down in 1962 during the Cuban missile crisis.* Its pilots called it the 'Dark Lady'.

It took several years to design and build the radiometer and install it on the plane. The first flight launched in July 1976, and by December the results were in, though they were not what had been expected (see Fig. 26). The dipole anisotropy was revealed, with an amplitude of ±0.0035 K, a little more than 0.1% or one part in 1000.[8] The radiation does indeed appear slightly warmer in the direction of Earth's travel, slightly cooler in the opposite direction. But the direction was wrong. It had been expected to be warmest in the direction of Aquarius, consistent with the direction of rotation of the Milky Way, and coolest in the direction of Leo. The opposite was true.

In addition to rotating, it transpired that the Milky Way is moving with a peculiar velocity of 600 km/s. This was completely unexpected. It would seem that the Milky Way (and, by inference, the Local Group of galaxies) is being drawn by gravity towards some unseen yet massive structure, at odds with the assumption of a smoother, more isotropic universe. 'The Universe may be much more inhomogeneous than we

---

* It also inspired the name of the Irish rock band U2.

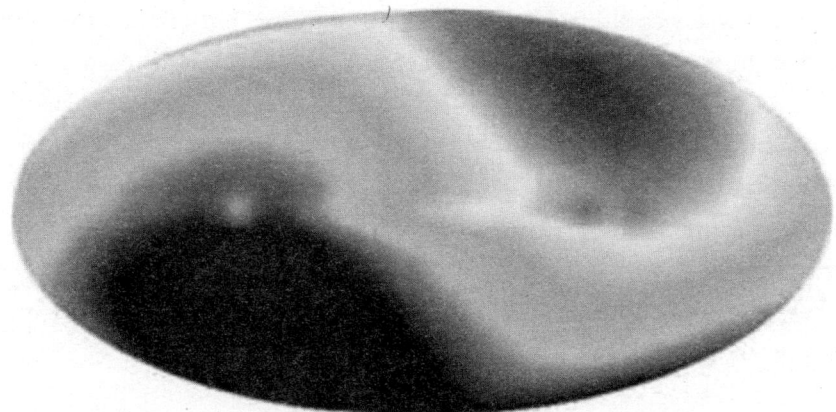

Fig. 26 The motion of the Milky Way galaxy through the cosmic background radiation causes a small shift in measured temperatures, the radiation appearing a little hotter as we move towards it, a little cooler as we move away, called the 'dipole anisotropy'. In this picture, hotter regions to the top right (with a temperature a little over 0.003 kelvin higher than average) appear dark grey, cooler regions to the bottom left (0.003 kelvin lower than average) appear black. The plane of the Milky Way lies horizontally along the centre.

have realized until now', Smoot and his colleagues concluded.[9] De Vaucouleurs would likely have agreed. In addition to the Milky Way and the Virgo cluster, the Virgo supercluster includes a group of galaxies in the constellation Leo.

Aside from the expected dipole anisotropy, there appeared to be no other discernible variation in temperature, at least to one part in 3000, or 0.03%. This posed a twofold puzzle. At the time of recombination, the universe was expected to be large enough for its different parts to be no longer connected. The limiting speed of light meant that distant parts of the universe were no longer in 'causal contact' and so could not influence each other. If the temperature of the cosmic background radiation happened to be different in these distant parts (and there was no good reason to suppose otherwise) then it would be impossible for these to come to thermal equilibrium. The stubborn uniformity of the temperature of the cosmic background (and, by inference, the universe as a whole) became known as the 'horizon problem'.

But there was also no good reason to suppose that the matter and radiation present in the universe at the moment of recombination was distributed strictly uniformly—the very existence of stars and galaxies told against this. And if it wasn't, there was every reason to expect that the temperature of the cosmic background radiation should indeed be different when we look in different directions across the sky. But it wasn't, at least to one part in 3000. This was admittedly still a factor of 3 larger than the predictions of the Harrison–Zeldovich spectrum, though it seems that Smoot was unaware of this result at the time.

Cosmic Inflation

Alan Guth almost didn't make it. In November 1978, Dicke was scheduled to give a short series of lectures at Cornell University in Ithaca, New York, in honour of Bethe. Guth, now on his third stint as a postdoc, was building a career as a particle theorist and had little interest in astrophysics or cosmology. He was also nursing a bout of bronchitis. But, just a month earlier, the Nobel prize committee had announced that Penzias and Wilson would receive the 1978 physics prize, and Guth was sufficiently intrigued to want to make the effort.

It was just as well that he did.

From Dicke, Guth learned about the flatness problem and the extraordinary fine-tuning that appeared to be required to produce our observable universe. He was fascinated. Later that same week he was introduced to another, seemingly unconnected, problem by a fellow Cornell postdoc, Henry Tye. Over lunch, Tye explained that theoretical structures developed to transcend the standard model of particle physics—called grand unified theories—might predict the existence of magnetic monopoles.

Guth had spent several years working on the theory of magnetic monopoles, hypothetical particles that produce magnetic fields in much the same way that charged particles produce electric fields. But whereas opposite electrical charges can be separated, opposite north and south magnetic poles cannot. You can't separate north from south poles by cutting a bar magnet in half. All you get are two smaller bar magnets. Guth was unfamiliar with grand unified theories, or GUTs. Tye brought him up to speed.

The standard model of particle physics actually consists of three distinct quantum field theories each dealing with a different kind of interaction involving a different force. We are most familiar with the electromagnetic force, established between electrically charged particles and 'carried' from one to the other by massless photons. We are likely less familiar with the weak force, involved in the process of beta radioactive decay, which turns a neutron into a proton, whilst ejecting a high-speed electron and a rather mysterious companion, an electron anti-neutrino.

A rash of discoveries at various particle accelerators had by the late 1960s suggested that protons and neutrons are, in fact, composite particles consisting of *quarks* which possess, in addition to oddly fractional electric charges, a further property called flavour—up and down. Not everybody in the particle physics community was convinced but, a decade later, quarks were formally accepted into the particle physicists' lexicon.

The weak force operates on quark flavour and is carried by massive, 'heavy' photons, called W (W^+ and W^-) and Z (neutral Z^0) particles. Finally, the strong force tethers the quarks together inside protons (containing two up, and one down quark) and neutrons (one up, two down quarks) and operates on another property of the quarks called colour—red, green, and blue—carried by massless particles called gluons which also possess a colour 'charge'. Strictly speaking, this colour force *is* the 'strong' force, as it spills beyond the boundaries of the protons and neutrons and serves to bind these together inside atomic nuclei.

Today, the standard model consists of three 'generations' of quarks with six different flavours—up, down, strange, charm, top, and bottom, differing only in their masses. These are matched by three generations of 'leptons'—the electron, muon, and tau, also with different masses, and their corresponding neutrinos. The force carriers include the photon, W and Z particles, and eight different kinds of gluon. To this we need to add one further particle.

As they stand, the individual quantum field theories are unconnected, and physicists cried out for some organizing principles that might explain their interrelationships. In the early 1940s it was realized that the weak force might be so very much weaker than electromagnetism because this force is carried by massive particles, and this confines the range over which the force can operate to within the boundaries of atomic nuclei. If the masses of the weak force carriers could be somehow reduced to zero, like the photon, then the weak force would in theory operate over larger distances and appear as strong as electromagnetism. Alternatively, if the electromagnetic and weak forces had once been unified in a single 'electro-weak' force, then something must have happened in the past to cause the weak force carriers to gain mass and so split the force into the two distinct forces we recognize today.

That 'something' is called the *Higgs mechanism*, devised in 1964 and named for English theorist Peter Higgs. In this mechanism, interaction with another quantum field called a 'Higgs field' breaks up the electro-weak force, carried by four massless particles, two charged ($+$ and $-$) and two neutral. The charged particles 'gain mass' in this interaction to become the massive W^+ and W^- particles, and the neutral particles mix together to produce the massive Z^0 and the massless photon.

In 1967, Steven Weinberg had made use of this mechanism to predict the masses of the W and Z particles. It says something about the confidence of the Nobel prize committee that Weinberg, American theorist Sheldon Glashow, and Pakistani theorist Abdus Salam were awarded the 1979 physics prize for their work on electro-weak unification, four years *before* the W and Z particles were actually discovered at CERN in Geneva (with masses extraordinarily close to Weinberg's predictions). This success sponsored a near 30-year hunt for the fundamental particle of the Higgs field, called the *Higgs boson*, which took its rightful place in the standard model in 2012.

Although these discoveries lay in the future, in 1978 the successful unification of the electromagnetic and weak forces led particle theorists to believe that further unification of the electro-weak and strong forces was possible, in a grand unified theory, leaving only gravity out in the cold. The first theory of this kind had been proposed by Glashow and Howard Georgi in 1974, but the formulation of GUTs was hampered by the simple fact that the energy at which the electro-weak and strong forces could be expected to become unified was extraordinarily large, corresponding to a temperature estimated to be about 10^{29} K. Not a problem for the very earliest moments following a Big Bang, perhaps, but substantially beyond the realm of human experience. The theorists were obliged to fly blind, speculate, and put their faith in the mathematics.

Aided by Tye, Guth set about analysing the fate of magnetic monopoles in the early universe, in essence attempting to deduce a 'divine curve' for the density of monopoles similar to those for matter and radiation. The GUTs predicted that monopoles were expected to be produced as the early universe expanded and cooled below 10^{29} K, undergoing a 'phase transition', presumed to be triggered by another hypothetical Higgs field, in which the unified force split into electro-weak and strong forces. Phase transitions, such as the condensation of water vapour to liquid, or the freezing of liquid water to ice, are very familiar. But it was novel for such physics to be applied to the entire universe. A subsequent phase transition would then split the electro-weak force into the electromagnetic and weak forces.

But the GUTs then available suggested that monopoles should be produced in great abundance in the first phase transition, contradicting their complete absence in today's universe. Guth needed to figure out how to suppress this production. By this time he had taken on another post-doc position—his fourth—and had moved from Cornell to the Stanford Linear Accelerator Center (SLAC) in California, where he hit on an idea. He figured that the monopoles could be suppressed if the universe had undergone a phase transition with a large amount of *supercooling*.

When liquid water condenses, impurities in the liquid or irregularities
on the surface of its container act as nuclei for the formation first of small
'seed' crystals, which grow into extended crystals which join together
to form the lattice structure we know as ice. But it is possible to cool
water that is free of impurities such that it remains liquid to temperatures up to 50 °C below the freezing point. This is supercooling. Translated to the early universe, bubbles of new phase could be expected to
form as the universe supercools below its first transition temperature of
10^{29} K. The bubbles then grow at nearly the speed of light before col-
liding and merging, eventually completing the phase transition. Guth
found that the original Glashow–Georgi GUT allowed for a relatively
calm and orderly rate of bubble formation, just what was required to
suppress the production of monopoles.

He told Tye the good news. As they worked on a draft paper, Tye suggested they check an important assumption. They had assumed that the
rate of expansion of the universe would be unaffected by supercooling.

Guth sat down in early December 1979 to figure out what supercooling might imply for the expansion of the universe. What he discovered
was quite shocking. It made sense to use the same kind of Higgs field
thought to be responsible for the splitting of the electro-weak force,
and its properties suggested that supercooling would trap the field in
what is known as a 'false vacuum', a temporary state characterized by an
enormous energy density which creates a strongly repulsive gravitational
field. Bubbles of 'true vacuum' would then form, expand, and merge
together.

The false vacuum acts similarly to Einstein's cosmological term,
except that it is much, much, larger. Lemaître had suggested that the cosmological constant could be considered to represent vacuum energy of
the order of 10^{-27} g/cm³, responsible for driving the expansion of the
universe. But the energy characteristic of the false vacuum was judged
to be typically 10^{80} g/cm³, sufficient to drive an extraordinary exponential expansion. The details were not sensitive to the nature of the GUT
used to calculate them. Guth estimated an expansion rate in which the
universe doubles in size every 10^{-37} s. 'I do not remember ever trying

to invent a name for this extraordinary phenomenon of exponential expansion', he later wrote, 'but my diary shows that by the end of December [1979] I had begun to call it *inflation*'.[10]

In an inflationary cosmological model, it makes no difference in principle how much mass–energy the early universe contained. It doesn't matter what the initial rate of expansion was. Early proponents of inflation theory argued that no matter what the shape of space prior to inflation, when inflation was done, *flat* space was the only possible result. Guth put an entry about this 'spectacular realization' in his notebook: 'This kind of supercooling can explain why the universe today is so incredibly flat—and therefore resolve the fine-tuning paradox pointed out by Bob Dicke'.[11]

Only in part. However it is achieved, and whatever the critical density ρ_c, a spatially flat universe still requires a density parameter $\Omega = \rho/\rho_c = 1$, and inflation could not bridge the gap between this and a density parameter for baryonic plus dark matter determined to be $\Omega_M = 0.2$.

But inflation also suggested itself as a solution to the horizon problem. Guth learned about this over lunch with colleagues in the SLAC cafeteria, just weeks after his spectacular realization. The argument goes that at the onset of inflation the universe was so small that every part of it was within reach of particles travelling at the speed of light—every part of the universe was within causal contact with every other part, and so was able to come to thermal equilibrium. Uniformity prevailed. So, when the universe supercooled and experienced an ultra-short burst of cosmic inflation, this uniformity was carried through to the larger universe when inflation was done.

A Graceful Exit

There is nearly always a 'but', and Guth now discovered a big one. This form of cosmic inflation had no graceful exit. Supercooled water will eventually form seed crystals which rapidly multiply and coalesce as the phase transition completes its journey to solid ice. But Guth's inflationary universe prevented this from happening. Although the bubbles of true vacuum would form and expand at near light-speed, the burst of extraordinary exponential expansion would carry them too far apart and prevent them from merging. Einstein's dictum concerning the limiting speed of light applies to objects moving through space, not to the expansion of space itself. The phase transition could never be completed, and the result would be an empty, 'cold Swiss-cheese' universe.

Guth set off on a long lecture tour of the US, risking that his work might be anticipated by another theorist and published before he could write it up and publish it himself. Sitting in the audience of Guth's lecture at Harvard on 5 March 1980, a young particle theorist called Paul Steinhardt was first exhilarated by what Guth had to say, then utterly dismayed at its ending. 'I simply could not believe that such a beautiful idea could fail so catastrophically'.[12] The result was hardly catastrophic for Guth himself. Interest in inflationary cosmology expanded with each lecture, and by the end of his tour he had gathered no less than eight formal and informal job offers. The eternal postdoc phase of his career was over. In the event, he accepted none of these offers, stuck his neck out, and pushed for a position at MIT, where he had studied for his PhD. His gamble paid off, and he was appointed as an associate professor.

Steinhardt was now hooked on inflation, and with Andreas Albrecht worked out how to avoid catastrophe. Russian theorist Andrei Linde arrived at the same solution around the same time. The properties of the Higgs field that had been adopted to split the electro-weak and strong forces were highly speculative, and there was no real reason to suppose that it should have broadly the same properties as the Higgs field used to split the electromagnetic and weak forces. So, instead of trapping the universe in a false vacuum, these theorists proposed a Higgs field with properties that allowed the universe to evolve a little more gracefully.

Linde called it 'new inflation'. It is also called 'slow-roll inflation'. In this scenario the false vacuum remains but the universe doesn't become 'trapped'. Instead, it evolves more slowly from its high-energy state.

Here, the word 'slow' is used advisedly. Inflation still happens extremely abruptly, within 10^{-35} to 10^{-32} s after the Big Bang, sufficient partially to resolve the flatness and horizon problems, whilst allowing the phase transition to complete, converting the energy stored in the Higgs field into hot matter and radiation. Now inflated, the universe continues its gentler expansion and cooling.

The Nuffield Workshop

Inflation was an idea whose time had come. As Albrecht and Steinhardt were writing up their paper on new inflation, Steinhardt met Michael Turner, an astrophysicist at the University of Chicago. They began to collaborate, and were soon fretting over the possibility that if new inflation solved the horizon problem, the resulting universe would be too smooth—too homogeneous and isotropic—and so incapable of forming stars and galaxies. They needed to find a way to introduce cosmic seeds.

The solution suggested itself and, to a certain extent, had even been anticipated by Lemaître. Quantum mechanics imposes constraints on the precision with which pairs of so-called 'conjugate' properties can be known. These properties are governed by an *uncertainty principle*, first elucidated in 1927 by German physicist Werner Heisenberg. Examples of such pairs are position and momentum (mass × velocity), and energy and the rate at which energy changes in time. One consequence of the uncertainty principle is that quantum fields are never still.

Think of it like this. Suppose we could measure the energy of a quantum field at some location in space with unlimited precision. Now suppose that the energy of the field at this location is fixed and unchanging. This would mean that we could establish both the precise energy and precisely how this energy is changing in time (or not, in this case), violating the uncertainty principle. Nature does not cooperate in this way. Instead, we find the greater the precision in energy, the greater the uncertainty in its rate of change in time. This does not depend on measurement—left to itself the energy of a quantum field will fluctuate randomly. Quantum fluctuations in an electromagnetic field were predicted in 1948, and verified (though with large experimental errors) 10 years later. This is the Casimir effect, named for Dutch theorist Hendrik Casimir. Experimental verification of Casimir's predictions to within 5% was achieved in 1997.

In the bigger picture of the large-scale universe, such fluctuations are unimportant. But the pre-inflationary universe was small enough for quantum effects to dominate. The Higgs field thought to be responsible for splitting the electro-weak and strong forces is just another quantum field, and would therefore be expected to experience quantum fluctuations. This was not in itself a revelation. But the fluctuations would mean that in some parts of the universe inflation might proceed a little faster and end prematurely, leading to an under-production of matter (an under-density). In other parts inflation might proceed more slowly and be prolonged, leading to a slight over-production of matter (an over-density). These small differences would have been frozen in place and stretched across the universe by inflation, producing precisely the cosmic seeds required to make stars and galaxies as the post-inflationary universe continued to evolve.

The solution suggested itself to other physicists at around the same time. At the University of Cambridge, Stephen Hawking concluded that the resulting fluctuations in matter density would be scale-invariant, consistent with the Harrison–Zeldovich spectrum. But the calculations were extremely tricky, involving several simplifying assumptions, and the theorists could not agree on the magnitude of the density fluctuations that would result. Steinhardt and Turner estimated 10^{-16}, much smaller than required. Hawking estimated 10^{-4}, and Guth estimated 10^4, much too strong. All were suspicious of each other's calculations.

In June and July 1982, Hawking and his collaborator at Cambridge, Gary Gibbons, organized a workshop on the very early universe, supported by funds provided by the Nuffield Foundation, a charitable trust

Photos 11–14 Participants at the Nuffield Workshop in June/July 1982 included (left to right) Alan Guth, Stephen Hawking, Paul Steinhardt, and Michael Turner.

set up in 1943 by William Morris (Lord Nuffield), the founder of Morris Motors. Steinhardt and Turner attended, as did James Bardeen, with whom they had collaborated on certain important aspects of their calculation. Guth also attended, as did Russian theorists Linde and Alexei Starobinsky.

The workshop lasted three weeks and the effort was intense, as the theorists worked long into the night to refine their calculations. Hawking was a highly respected theorist but not yet a public figure (his highly successful popular book *A Brief History of Time* was first published in 1988). His motor neurone disease had already confined him to a wheelchair but he could still speak unaided, though his speech was slurred and his lectures required an interpreter.

By the end of the workshop, they had reached a consensus. The fluctuations in matter density were scale-invariant, as required, but their magnitude was sensitively dependent on the choice of the underlying GUT. The popular Glashow–Georgi GUT predicted density fluctuations of the order of 100, a million times larger than required and clearly in disagreement with the evidence from studies of the cosmic background.[13] The result would be a universe full of black holes.

This was disappointing but, in truth, in the early 1980s GUTs were very speculative. The Glashow–Georgi GUT allowed transformations between quarks and leptons, which meant that a quark inside a proton could in principle transform into a lepton. 'And then I realized that this made the proton, the basic building block of the atom, unstable', Georgi said. 'At that point I became very depressed and went to bed'.[14] A search for evidence of radioactive protons later came up empty, and enthusiasm for GUTs had all but evaporated by the end of the decade.

The theorists realized that they had been trying to drive inflation with Higgs fields that had been designed for very different purposes. Steinhardt and Turner turned the problem on its head. They prescribed a set of rules for viable candidates for the quantum field, now called the 'inflaton' field—the field that drives inflation—to distinguish it from the Higgs fields used by particle theorists. Such candidates were designed to produce fluctuations of the right order of magnitude. The aim was not to use theory to predict the fluctuations, but simply to show that the theory could be engineered to be compatible with them.

This was about as far as they could go. Inflation could provide part of the explanation for why the universe is flat and why the temperature of the cosmic background radiation is so uniform, whilst allowing the possibility that quantum fluctuations in the inflaton field, amplified by inflation, could account for its large-scale structure. Little progress could be made until more detailed observations became available.

Challenger

In 1975, the European Space Agency (ESA) joined the project to build the large space telescope, offering 15% of the funding in exchange for a guarantee from NASA that 15% of the telescope time would be awarded to European astronomers. Congress agreed funding of $36 million in 1977, and work on the design of the telescope was begun in earnest the following year. At a very early stage it was understood that the telescope would be carried into orbit and routinely serviced by a fundamentally new type of reusable, low-Earth orbital spacecraft, called the space shuttle. This was conceived in the late 1960s and developed during the 1970s. The space shuttle *Columbia* completed its maiden voyage in April 1981.

That year, NASA chose to establish the Space Telescope Science Institute (STScI) at Johns Hopkins University in Baltimore, Maryland, charged with the tasks of providing long-term guidance for the large space telescope project and engaging with astronomers around the world. Riccardo Giacconi, a pioneer of space-based and X-ray astronomy, was appointed as the Institute's first director. He set about recruiting a 250-strong staff. In 1983, the telescope was renamed as the Hubble Space Telescope (hereafter HST). It was scheduled to be launched later that year, but was overtaken by technical delays and budget problems. The launch date slipped to October 1984, then April 1985, then March 1986, then September 1986. Project costs escalated to over a billion dollars.

On Giacconi's urging, NASA and STScI convened four scientific panels to scope out 'Key Projects' that would be allocated blocks of telescope time. The draft of a proposal for a key project on the Hubble constant was developed in 1984 by a team of 13 astronomers led by Marc Aaronson, based at the University of Arizona's Steward Observatory. The team included Canadian-American astronomer Wendy Freedman, acting as Aaronson's deputy, then at the Carnegie Observatories in Pasadena, California, American Robert Kennicutt, Australian Jeremy Mould, Barry Madore, and John Huchra. Madore had acted as Freedman's thesis advisor at the University of Toronto, and by now they were married. The Key Project Team awaited the outcome of peer review as the launch date approached.

Smoot was also waiting. He had begun commuting between Berkeley in California and NASA's Goddard Space Flight Center in Maryland and faced the challenge of demonstrating that the DMR would work as expected in orbit aboard a satellite. He was briefly knocked off balance in 1980 by announcements from two groups—one led by Wilkinson and the other by Francesco Melchiorri at the University of Florence—that they had detected temperature variations in the cosmic background consistent with a quadrupole signature. Whereas the dipole pattern has two poles, a quadrupole pattern has four, two hot and two cold. No such pattern had been found in the experiments flown aboard the U-2 spy planes.

Both Wilkinson and Melchiorri had carried out balloon-borne experiments. Although Smoot pushed for further U-2 missions, extensive delays meant he had little choice but to overcome his reluctance and return to ballooning. Two flights from the US National Scientific Ballooning Facility in Palestine, Texas, went without a hitch. The dipole was evident, but there was no quadrupole. Smoot switched attention to the southern sky, but then disaster struck once again. A balloon launched in November 1982 from the Instituto Nacional de Pesquisas Espaciais in Brazil failed to release its payload and sailed off into the sunset, carrying the data with it. Remarkably, the missing balloon was found over a year later, in January 1984, in a forest preserve more than a hundred miles from its expected landfall. Pieces of equipment were recovered from display in a local bar, and hasty negotiations with the poacher who had found it recovered the rest. Even more remarkably, the cassette tape carrying the data was still readable. No quadrupole.

By the mid-1980s the sensitivity of the DMR was approaching the Harrison–Zeldovich prediction of 1 part in 10,000, but no further temperature anisotropy could be found beyond the dipole. All hopes were now pinned on the COBE satellite mission, which NASA had by now approved, with one significant modification. The satellite was originally intended to launch aboard a Delta rocket, but NASA now preferred to configure it for launch aboard a space shuttle, scheduled for late 1988.

On 28 January 1986, the space shuttle *Challenger* was prepared for launch from the Kennedy Space Center in Florida. This was to be *Challenger*'s tenth orbital flight, and the 25th space shuttle mission. It was freezing cold. The overnight temperature had been predicted to fall to −8 °C, rising to −3 °C at the scheduled launch time of 9:38 a.m.

Engineers at Morton Thiokol, manufacturer of the solid rocket boosters used to lift the shuttle for the first two minutes of its flight before being jettisoned, expressed concerns about the effects of such low temperatures on the elasticity of 'O'-ring seals used in field joints between booster sections, so called because these joints are sealed 'in the field' just prior to each flight. They recommended that the launch be delayed until the air temperature rose above 12 °C.

They were overruled, and the decision was taken to proceed with the launch.

Challenger launched at 11:38 a.m., and exploded 73 seconds later. The crew, captained by Francis Scobee, included pilot Michael J. Smith and mission specialists Ellison Onizuka, Judith Resnik, and Ronald McNair. Payload specialists, trained to manage a specific shuttle payload, included Gregory Jarvis and schoolteacher Christa McAuliffe, the first (and only) participant of the Teacher in Space Project announced by President Ronald Reagan in 1984.* All were killed. Sitting on the committee subsequently convened to investigate the fatal accident, American physicist Richard Feynman learned of the reservations of the Morton Thiokol engineers. In a public meeting of the committee, he demonstrated what would happen to the elasticity of an 'O'-ring seal when dipped in a glass of iced water: '... for more than a few seconds, there is no resilience in this particular material ... I believe that has some significance for our problem'.[15]

Smoot vividly remembered the *Challenger* disaster. 'I was stunned', he later wrote. 'We all were. We grieved for the astronauts. The tragedy of the accident was uppermost, but slowly the probable implications for COBE began to dawn'.[16] NASA suspended all shuttle flights indefinitely.

---

* The Teacher in Space Project was cancelled in 1990, but teacher Barbara Morgan (who was back-up to McAuliffe in 1986) was a Mission Specialist aboard the space shuttle *Endeavour* in August 2007. That year NASA introduced its Educator Astronaut programme, in which qualified teachers were selected and trained as full-time astronauts.

The launch of the HST, which had already been delayed for over three years, was also affected. But the planning continued. Later in 1986, NASA awarded the HST Key Project on the Extragalactic Distance Scale to the team led by Aaronson. Huchra explained that its primary goal was to 'beat down the errors... to derive a value of the Hubble Constant good to 10 per cent'.[17] Alas, the tragedy was not yet done. Aaronson was killed in a freak accident in April 1987 at the Kitt Peak National Observatory's Mayall Telescope. He was just 37 years old. Freedman acknowledged that 'It was an awful time'.[18] Freedman and Madore's first child was born just a week later.

The Great Wall

The Harvard–Smithsonian Center for Astrophysics (CfA) was formed in July 1973 as a collaboration of the Harvard College Observatory and the Smithsonian Astrophysical Observatory. This is a collaboration bound by a memorandum of understanding. It was not a merger. Both institutions retained their separate identities and continue to operate independently, but their astrophysical research efforts are coordinated by the CfA, housed on the Harvard campus under a single director. The collaboration was born partly in response to the Harvard Observatory's waning fortunes. Many in the research community judged that the glory days associated with Pickering and Shapley were now firmly behind it. Although the Smithsonian was better funded and resourced, to the chagrin of Smithsonian astronomers, when news media reported on breakthroughs led by 'Harvard astronomers' it was clear which institution was getting the credit.

Marc Davis arrived to take up an assistant professorship at Harvard in the autumn of 1974. He had gained his PhD at Princeton and had left with Peebles' sage advice ringing in his ears. There was more to the universe than two numbers. It appeared to have a large-scale structure that was poorly understood. Following further discussions with Smithsonian astronomer David Latham, Davis decided that what was needed was an ambitious survey of galaxy redshifts.

Despite the continuing uncertainty and anguished debate about the size of the Hubble constant, and what seemed a readily available source of new data from redshift measurements, the catalogue of galaxy red-shifts had not expanded beyond that published by Humason, Mayall, and Sandage in 1956. Redshift is measured from the spectrum of light dispersed from a distant galaxy, and gathering enough light from such a faint source to take a reliable spectrum demanded too much telescope time. Davis became fixated on the ambition to measure the redshifts of all 2000 galaxies of magnitude 14.5 or brighter observable in the north-ern sky, using the Smithsonian's 60-inch telescope at Mount Hopkins Observatory in Arizona.

Davis and Latham were joined in 1976 by Huchra and a young grad-uate student called John Tonry. They built a new spectrograph that could amplify and record the spectrum electronically and compare it with a reference spectrum. Tonry wrote a computer program that would automatically calculate the redshift. They called it the Z-machine. With everything working as it should, they could measure between 10 and 15 redshifts each night.

By 1981, the survey of 2400 galaxies was complete. The results were striking. 'The space distribution of galaxies is frothy, characterized by large filamentary superclusters up to 60 Mpc in extent, and correspond-ing large holes devoid of galaxies'.[19]

In the meantime, Davis had come up for promotion but Harvard had prevaricated. By the time George Field, then director of the CfA, offered him a tenured position, Davis had already decided to move to Berke-ley. With Margaret Geller, a former student of Peebles recently returned to Harvard from Cambridge in England, and Geller's graduate student Valérie de Lapparent, Huchra extended the survey to magnitude 15.5. 'The galaxies appear to be on the surfaces of bubble-like structures ... [which] have a typical diameter of $\sim 25h^{-1}$ Mpc. (Here h is a dimension-less form of the Hubble constant defined as $h = H_0/100$ km/s/Mpc.)

The largest bubble in the survey has a diameter of $\sim 50h^{-1}$ Mpc'.[20] A Hubble constant $H_0 = 50$ km/s/Mpc gives $h = 0.5$, a typical bubble diameter of 50 Mpc and the largest bubble diameter of 100 Mpc. These sizes were reported in terms of the dimensionless h because of the uncertainty in the Hubble constant.

A map of a 6° slice of the sky in the direction of Virgo and Coma, containing 1057 galaxies, featured what looks like a stick-man structure, Fig. 27(a), which became an iconic image. It looks like a child's drawing, but extends over 500 million light-years, with a body comprising hundreds of galaxies in the Coma cluster and arms of thin sheets of galaxies. Interpretation requires some care. The Coma cluster is spherical but this is a map of redshift velocity ($v = cz$), and the collective gravitational pull of the cluster results in a spread of local velocities which stretches the redshifts along the line of sight.

'The striking thing you see', Geller told *Time* magazine in January 1986, 'is the vast empty regions surrounded by a thin structure'.[21] She compared it to soap suds in the kitchen sink.

By 1989, the redshift survey had extended to include eight 6° slices and 5800 galaxies. Slices at different declinations—tilted slightly from the plane of the slice shown in Fig. 27(a)—showed that the filaments extending from the arms of the stick-man structure form a vast sheet of galaxies which Geller and Huchra called the 'Great Wall', illustrated in Fig. 27(b). It has dimensions of 60–120 Mpc × 170–340 Mpc, depending on the value of the Hubble constant. It was a sobering result. '... the size of the largest structures we detect is limited only by the extent of the survey'.[22] A more extensive survey would likely reveal even larger structures.

The redshift survey exposed a universe that the theorists could not account for. Theories devised to describe large-scale structure produced patterns that did not fit, and voids that were not big enough. The age problem persisted. The universe did not appear old enough for such elaborate structures to have formed from more homogeneous and isotropic beginnings in the time available to it. Inflation was supposed to have

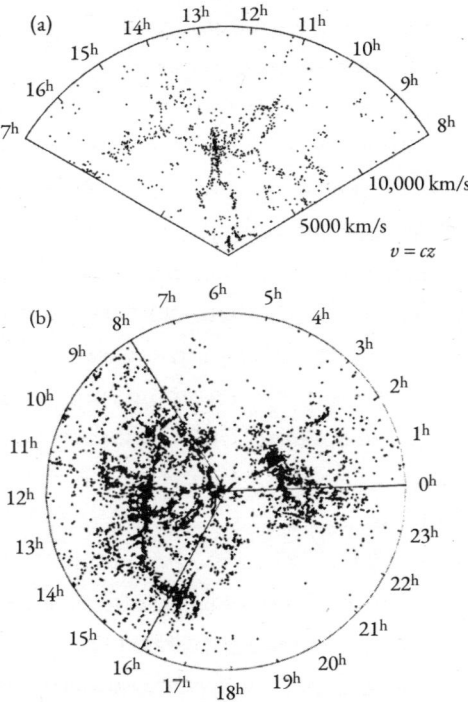

Fig. 27 The second CfA galaxy redshift survey produced maps of 6° slices of the universe. The slice in (a) lies in the direction of Virgo and Coma and features a stick-man structure extending over 500 million light-years. The filaments stretching from the arms of the stick man are part of a vast sheet of galaxies which Geller and Huchra called the 'Great Wall', shown on the left in (b). The Perseus–Pisces chain appears on the right.

solved the flatness problem but flatness demanded $\Omega = 1$ and the distribution of redshifts was firmly consistent with $\Omega = 0.2$, with dark matter accounting for 80% of this density.

The theorists understood that the formation of large-scale structure is intimately linked with the nature of dark matter, categorized as 'cold', 'warm', or 'hot'. This was not so much about temperature, but about the speeds of dark matter particles (whatever they are) in the early universe, as determined by the distance (or length) that particles travel freely through space without being scattered. If the *free-streaming length* is shorter than the size of a protogalaxy (one that will later evolve into a dwarf galaxy), then the dark matter is 'cold'. It is warm if the free-streaming length is about the size of a protogalaxy and 'hot' if it is much longer. It was becoming apparent that the structures visible in the red-shift survey told of a universe built from the bottom up: small objects forming first and merging to form larger and larger structures. Only cold dark matter would allow this.

But there was still no telling what cold dark matter is, and the long sought after anisotropies in the cosmic background remained elusive, calling into doubt the possibility of forming even small objects.

It was too soon to call it a crisis, but inflationary Big Bang cosmology was feeling the strain. In August 1990, Halton Arp, Geoffrey Burbidge, Fred Hoyle, Jayant Narlikar, and Chandra Wickramasinghe published a polemic in *Nature* pointing out its shortcomings. They wrote: 'The currently popular cosmological model is subject to many doubts based on observational data which suggest that, perhaps, there never was a Big Bang'. They proposed the steady-state model as an alternative with which the Big Bang cosmology 'fails to compete', reminding readers that '... it is undesirable to depend crucially on what is unobservable to explain what is observable'.[23]

8

Dark Energy and the Accelerating Universe

The uncertainty created by the *Challenger* disaster proved to be relatively brief. Although the space shuttles were grounded, there remained other ways to get satellites into orbit, and the COBE project team was now required to move extra quickly. There were enough parts left abandoned in old sheds for McDonnell-Douglas to build another few Delta rockets, and NASA gave the go-ahead for a COBE launch in early 1989, a seemingly impossible deadline. The scientists and engineers now scrambled to fit a satellite designed for a space shuttle into the much smaller rocket.

Another hiccup. In 1987 a joint Japanese–American team led by Toshio Matsumoto at Nagoya University and Andrew Lange and Paul Richards at Berkeley reported the results of measurements of the temperature spectrum of the cosmic background made by instruments aboard a sub-orbital rocket. They had found that the spectrum was not that of a true black body, causing tremendous excitement and once again calling the entire Big Bang cosmology into question.[1] Smoot was chastened. 'It added an urgency to our work, and suddenly NASA's early-1989 launch date, which had once seemed unrealistically early, now looked distressingly late'.[2]

Smoot experienced both elation and sadness when COBE finally left its clean room at the Goddard Space Flight Center to begin its journey west to Vandenburg Air Force Base in California. The satellite was successfully placed in Sun-synchronous orbit on 18 November 1989.

At a meeting of the American Astronomical Society in Arlington, Virginia, just a few months later in January 1990, Mather explained that the mission was proceeding smoothly and would take another year or more to map the cosmic background across the entire sky. But he was able to present preliminary results from the Far InfraRed Absolute Spectrophotometer (FIRAS) instrument, based on just nine minutes of data-gathering, pointing towards the north galactic pole.

The COBE team had written these results up for publication in the *Astrophysical Journal* and Mather and Smoot had mailed the paper to the editor that morning. The results were published in May. They show a black-body spectrum with a temperature of 2.735 ± 0.06 K (Fig. 28).[3] Wilkinson presented the same results to a colloquium at Princeton on the same day. Both audiences broke into spontaneous,

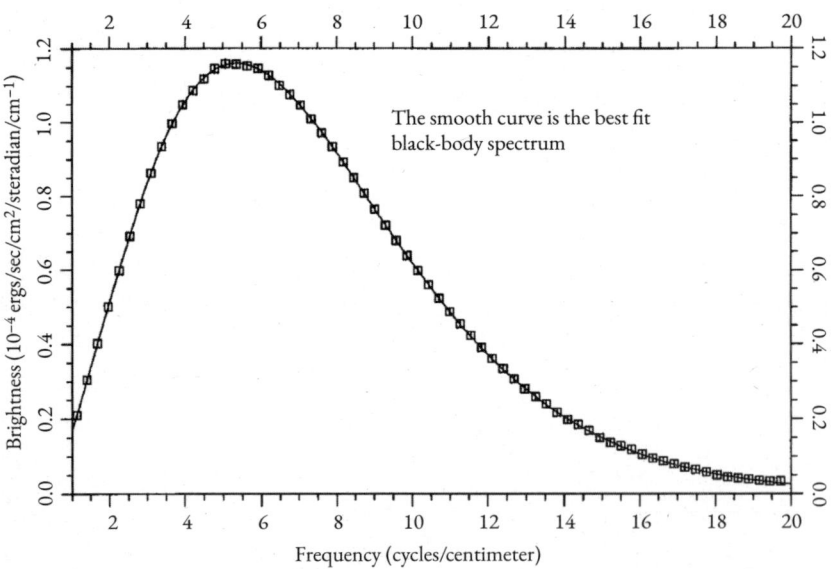

Fig. 28 Preliminary data gathered in just nine minutes by the FIRAS instrument aboard COBE confirmed that the cosmic background has a black-body spectrum with a temperature of 2.735 ± 0.06 K.

rapturous applause.[4] There was no discernible deviation from Planck's black-body radiation law. The applause was more an expression of relief than anything else.

But Smoot had nothing to add about the much-anticipated anisotropies. He was waiting for sufficient data to be gathered and analysed, and sought to maintain absolute secrecy as tensions began to build, both within the team and in the wider communities of astronomers and cosmologists. All he could admit was that any fluctuations in the temperature of the cosmic background had to be less than several parts in 100,000. By October 1991, some within the DMR group were arguing for publication of the tentative evidence they had gained. Others urged caution: mistakes had been made all too frequently, and all too recently. Smoot preferred to wait. He decided that they had to make one last check. They needed to rule out radio interference from the Milky Way.

Although he judged the chances were very slight that such interference could be responsible for the patterns in the temperature data they were now seeing, he couldn't rule it out. The available map of radio noise from the Milky Way was patchy and unhelpful. He felt he had no alternative but to make a map of his own, from radio telescope measurements at one of the few places on Earth where he could be sure to obtain good data: an observing station in Antarctica, just a few kilometres from the South Pole.

It was an inhospitable place. Visiting scientists learned to stack beer three cases high—the bottom case would freeze, and the middle case would be too cold to drink. But the case on the top would be fine. Smoot and his colleague Giovanni de Amici spent a month there in November/December 1991. Progress was far from smooth, but the new map showed that galactic radio emissions were not responsible for the patterns they could see in the temperature data obtained by the DMR aboard COBE.

Like Seeing God

More frantic analysis followed and, on 23 April 1992, Smoot stood at a podium at the annual meeting of the American Physical Society in Washington, DC. 'I've a lot to say, and not much time, so I'll get right to it'.[5] The anisotropy had been found (Fig. 29), with a magnitude a little more than 0.001%, or 1 part in 100,000, corresponding to temperature fluctuations measured in microkelvins. The features were scale-invariant 'in accord with the Harrison–Zeldovich ... spectrum predicted by models of inflationary cosmology'.[6] At a press conference organized after his presentation, Smoot struggled to find superlatives to convey the importance of the discovery. 'If you're religious', he said, 'it's like seeing God'.[7]

Mather and Smoot were jointly awarded the 2006 Nobel prize for physics. Ironically, Mather earned his share of the prize for showing that the temperature of the cosmic background is smooth and homogeneous, consistent with a thermalized black body conforming to Planck's radiation law. Smoot earned his share by showing that the cosmic background is not so homogeneous as to exclude the possibility of stars and galaxies.

The cosmology now seemed more settled and less ambiguous. A short burst of inflation was followed by frantic periods of particle creation as the Einstein–de Sitter universe ($\Omega = 1$, no cosmological term) continued to evolve with a gentler rate of expansion set to decelerate fully after an infinite amount of time. About 400,000 years after the Big Bang, the universe had expanded and cooled sufficiently for recombination to occur, releasing the cosmic background radiation.

But recombination occurs more slowly in regions of high matter density, and the radiation is therefore a little hotter there. These regions of matter over-density would be the seeds for formation of stars and galaxies as the universe evolved. The temperature variations are like a bloody thumbprint left at a cosmic crime scene, the signature of quantum fluctuations impressed on the large-scale structure of the universe by inflation.

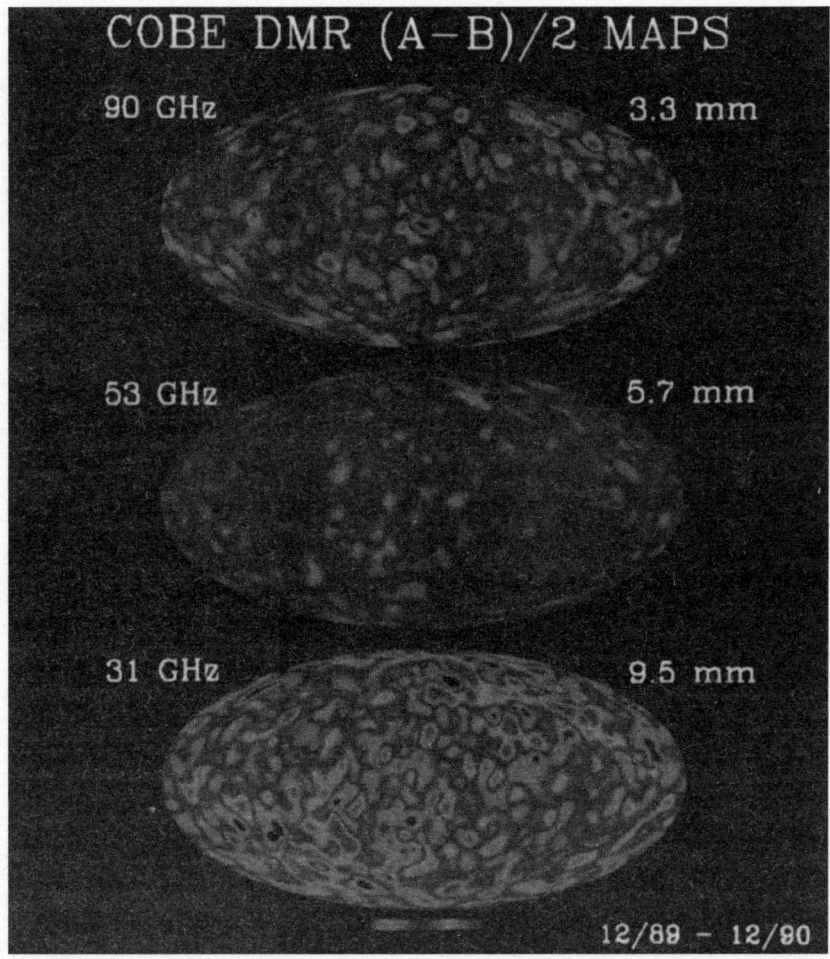

Fig. 29 These all-sky maps measured by the DMR instrument aboard the COBE satellite correspond to three different microwave frequencies (or wavelengths). All show small variations in temperature, the legacy of quantum fluctuations amplified by inflation which will act as seeds for star and galaxy formation. Darker areas correspond to colder temperatures, brighter areas to hotter temperatures.

The COBE results were hailed as a triumphant confirmation of inflationary Big Bang cosmology. But there remained some mysteries. No amount of theoretical speculation could resolve the puzzle of cold dark matter. And cosmologists continued to grumble about the size of the

Photos 15 and 16 John Mather (left) and George Smoot (right) were jointly awarded the Nobel physics prize in 2006.

measured density parameter, Ω. Even when dark matter was included, it was just too low.

Although not judged worthy of Nobel prize-level recognition, results from the third instrument aboard COBE—the Diffuse Infrared Background Experiment—were also significant. It detected far-infrared radiation from a number of early galaxies that had not been surveyed by IRAS, it enabled the origin of interplanetary dust to be traced to asteroids, and its data were used to build a model of the Milky Way's galactic disc.

A $2 Billion Lemon

COBE had been a great success, but the HST did not fare so well. The space shuttle programme recommenced in September 1988, and the HST was finally launched by the space shuttle *Discovery* in April 1990.

But within weeks it was obvious that something was wrong. Images from the telescope were sharper than any images available from ground-based telescopes, but they were not sharp enough. It was discovered that the HST's primary mirror had been precisely polished to the wrong shape, causing substantial spherical aberration. The HST's optics were perfect, but they turned out to be 'perfectly wrong'. Other systems and components began to fail.

The HST became an object of public ridicule, with some declaring it a '$2.1 billion lemon'.[8] The American cartoonist Gary Larson published a Far Side cartoon featuring a blurry image of aliens waving mischievously from the cockpit of a flying saucer: 'Another photograph from the Hubble telescope'. Many scientists feared that NASA or the US Congress would simply write it off. The attentions of NASA and STScI scientists and engineers turned to the first servicing mission, scheduled for 1993, and tried to work out what could be salvaged. There was no question of replacing the mirror in orbit, or bringing the HST back to Earth for a refit.

But the precision with which the wrong shape had been polished suggested its own solution. The imaging problems had been picked up by the Wide Field and Planetary Camera (WF/PC) fitted aboard the telescope. NASA had decided to build a second instrument (WF/PC-2) as a back-up, and this was in the early stages of construction at the Jet Propulsion Laboratory in Pasadena at the time of the launch. John Trauger, the Principal Investigator for the new instrument, realized that the performance of the HST could be restored by fitting it with optics designed to produce aberration equal and opposite to that caused by the mirror. 'It would be like putting contact lenses on [the] Space Telescope, correcting its blurry vision of the Universe'.[9]

The design and installation of the correcting optics was overseen by John MacKenty, a senior scientist at STScI, and the servicing mission was conducted by astronauts aboard the space shuttle *Discovery* in December 1993. It involved five separate 'space-walks' (extra-vehicular activities or

EVAs). Tests conducted over the next six months demonstrated that the mission had been entirely successful (see Fig. 30). Its scientific value was quickly realized. In May 1994, astronomers reported on HST images of M87, a supergiant elliptical galaxy in the constellation Virgo, concluding that a disc of infalling ionized gas '... is strong evidence for a massive black hole in the center of M87'.[10]

Following Aaronson's tragic death, Freedman, Mould, and Kennicut had acted as co-Principal Investigators on the Key Project Team. The team would refer to them as the 'triumvirate'.[11] By October 1994, the team was reporting HST measurements of the distance to M100, one of the largest and brightest galaxies in the Virgo cluster, based on observations of Cepheid variables. A distance of 17.1 ± 1.8 Mpc implied a Hubble constant $H_0 = 80 \pm 17$ km/s/Mpc.[12] The team had yet to beat the errors down to the desired level, but this was a sign of things to come.

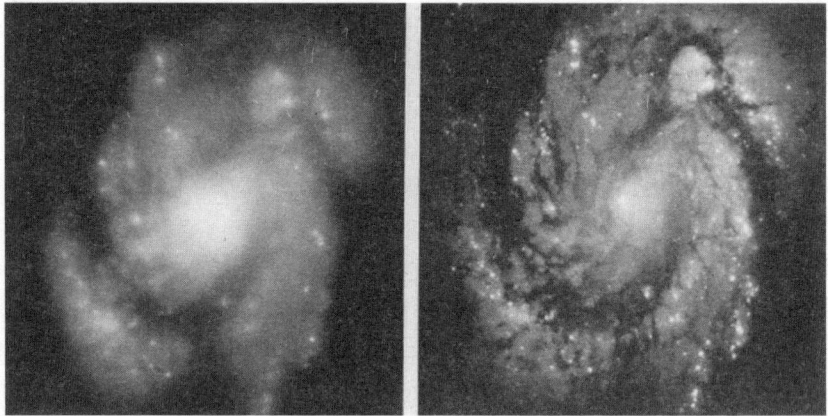

Fig. 30 The 'raw' image of the spiral galaxy M100 on the left was taken with the Wide Field and Planetary Camera (WF/PC) aboard the HST on 27 November 1993, just prior to the servicing mission. It shows the extent of blurring caused by spherical aberration of HST's primary mirror. The equivalent raw image on the right shows the same galaxy photographed with WF/PC-2, fitted with suitable correcting optics, on 31 December 1993.

Searching for Supernovae

The second CfA redshift survey reached to galaxies with recession speeds up to about 15,000 km/s. But the distances of these galaxies were largely unknown. The Hubble–Lemaître law $v = H_0 d$ could be used to infer distances up to about 200 Mpc (using $H_0 = 80$ km/s/Mpc). But unless some means could be found to determine the absolute magnitudes of the galaxies and hence (from the distance modulus) their distances, the redshift data could not be used to provide an independent determination of the Hubble constant. Perhaps more to the point, if the absolute magnitudes of the faintest, most distant, galaxies could somehow be measured, their corresponding look-back times (see Appendix 3) offered the possibility of glimpsing the universe at much earlier stages of its expansion history, and determining the second of Sandage's two numbers, the deceleration parameter.

There was a potential way out. When a star goes supernova it produces enough light to brighten an entire galaxy. If some way could be found to turn supernovae into standard candles, then measurements of their absolute magnitudes could be used to infer their distances, not just to the stars that had exploded but to their host galaxies too.

There were a few stumbling blocks. Not all supernovae are the same. In 1941 Rudolph Minkowski had identified at least two types depending on the absence (Type I) or presence (Type II) of spectral lines associated with hydrogen. Zwicky identified Types III, IV, and V, but these classifications are not in use today. Instead, astronomers have subdivided Types I and II based on their spectra and 'light-curves', the variation in time of the intensity of the light emitted, which rises swiftly to a peak and then declines typically over 100 days or so.

Type Ia supernovae exhibit a strong spectral line associated with silicon and broadly predictable light curves. This predictability arises from the nature of the physics involved. These are the signatures of thermal runaway within a binary star system, in which two stars are gravitationally bound to each other. One of the stars ages and eventually collapses to a white dwarf with a core consisting mostly of carbon and oxygen. As the second star matures to its red-giant phase, the white dwarf may drag matter from its neighbour's bloated outer layers, so building itself up once more. When the mass of the dwarf exceeds precisely 1.44 times the mass of the Sun, the conditions are right to trigger carbon burning in its core. But this is all too much for the hapless dwarf. The energy from carbon burning is released too quickly and the re-born star is torn apart in a supernova. The resulting shockwave can travel at speeds up to 20,000 km/s.

This is a scenario with highly predictable consequences. The dwarf has a predictable chemical composition and cannot go supernova until it accumulates a sufficient (and precisely determined) mass that gets it across the threshold for carbon burning. Consequently, the light emitted in the explosion has a characteristic luminosity and light-curve. Attention therefore turned to Type Ia supernovae as potential standard candles.

But then there was a further stumbling block, of literally astronomical proportions. Although supernovae explosions are happening all the time in galaxies across the universe, there is no telling where or when the next one will occur. To observe a Type Ia supernova, astronomers had to be looking in just the right direction at just the right moment. And supernovae are not stars or galaxies, which persist for billions of years—they are ephemeral. They fade. There is a limited time window of about three months in which to make observations.

Astronomers were generally sceptical. There were examples of 'peculiar' Type Ia supernovae that cast considerable doubt on the assumption of standard and predictable behaviour. Once a supernova had been spotted, hopefully very early in its fleeting existence whilst it is still brightening, a telescope was needed to find it and measure its light-curve and spectrum. Astronomer Robert Kirshner commented: 'If a supernova goes off once in a century per galaxy, that's roughly once in 5000 weeks, so if you want to see a nice fresh supernova at its brightest tonight you need to examine several thousand galaxies'.[13] This was not something astronomers felt inclined to do.

Systematically searching for supernovae was not a new idea, but recent developments in technology suggested ways in which this might be done efficiently. Telescopes could now be automated to scan large swathes of the night sky, 'wide-field' cameras could be used to capture images of tens-to-hundreds of galaxies at a time, and using a digital charge-coupled device (CCD) in place of the analogue photographic plate offered an enormous increase in sensitivity. The resulting digital images could now be compared not by eye, as Leavitt had done to find variable stars, but by computer image subtraction, pixel by pixel. A residual bright spot left after subtraction could be easily identified and then inspected to determine its origin and affirm (or not) that it was indeed a supernova. Once identified, the supernova's spectrum and light-curve could be measured.

Stirling Colgate was a Colgate family heir and expert on thermonuclear weapons. He had been summarily graduated from the Los Alamos Ranch School in New Mexico in 1942 when it was closed down at short notice. On the suggestion of J. Robert Oppenheimer, the school was identified as 'Site Y', the remote but accessible location for a new laboratory dedicated to atomic weapons research as part of the Manhattan project.[14] The site became the Los Alamos National Laboratory, and Oppenheimer became its first director.

Colgate had begun a search for supernovae in the late 1970s before the necessary technology was sufficiently developed. Alvarez read about his exploits in a magazine. He understood that Colgate was giving up on the project and suggested to Richard Muller (a former PhD student) that this was something that might be of interest. Muller enrolled his postdoc Carlton Pennypacker and a few graduate students, and established the Berkeley Automatic Supernova Search team. The team found its first supernova in May 1986, and recruited postdoc Saul Perlmutter shortly afterwards.

A few further supernovae had followed, but none exhibited a cosmologically significant redshift. The team had initially estimated that it would find 100 supernovae per year, but had come up with only a handful. Funding hung by a thread until 1988, when the Lawrence Berkeley Laboratory (LBL)* and the University of California at Berkeley were awarded substantial funding by the US National Science Foundation to establish a new Center for Particle Astrophysics (CfPA). The principal aim of the new centre was to explore potential particle origins for dark matter. But under its umbrella the supernovae search could continue, on probation.

The Berkeley team was not alone. Through the mid-1980s a group of Danish astronomers, Hans Nørgaard-Nielsen, Lief Hansen, and Henning Jørgensen at the Danish Space Research Institute and Copenhagen University Observatory, had trekked to the European Southern Observatory in northern Chile in search of Type Ia supernovae in galaxy clusters. They were rewarded on 9 August 1988. British astronomers Richard Ellis at the University of Durham and Warrick Couch at the Anglo-Australian Telescope measured its spectrum. The supernova set a new redshift record of $z = 0.31$, corresponding to a recession speed of about 93,000 km/s and a look-back time of about 3.4 billion years.[15]

Alas, the pioneering Danes were ahead of their time and, like Colgate, they abandoned the project after a couple of years of searching. They had the right ideas but their instrumentation was not quite up to the task. They were unable to search sufficiently large areas of the sky and their rate of discovery was consequently too low. Telescope time is allocated months in advance, and astronomers want to make the most of the time they have available. An urgent request from other astronomers to find and take the spectrum of an unlooked-for and unexpected supernova is unlikely to be greeted with much enthusiasm.

The Danish failure was sobering, but the Berkeley team could take heart. The Danes had, after all, succeeded in finding a Type Ia supernova (and, it seems, had found a second that they did not report). The Berkeley team had access to a much larger camera and a larger telescope.

---

* This became the Lawrence Berkeley National Laboratory in 1995.

But Pennypacker judged that the project needed an insurance policy. He recruited Gerson Goldhaber, a particle physicist who had been at the centre of the discovery of the J/ψ meson (a composite particle consisting of charm and anti-charm quarks) in November 1974. The discovery had put the charm quark firmly on the map, and entered particle physics folklore as the 'November revolution'. Pennypacker figured that any project in which Goldhaber was involved would be difficult for the CfPA to shut down.

In March 1992, Perlmutter used the Isaac Newton Telescope on La Palma in the Canary Islands to survey a patch of sky containing about 200 high-redshift galaxies. In late April and early May, Pennypacker surveyed the same fields, sending the images back to Perlmutter in Berkeley. Image analysis revealed a candidate supernova, SN 1992bi.* They now reached out to astronomers with a request to find the supernova and record its spectrum. Twelve attempts were made at four observatories. All failed because of poor weather conditions or equipment malfunction. Finally, on 29 August 1992, the spectrum was recorded at the William Herschel Telescope on La Palma. It set a new supernova redshift record, $z = 0.458$, corresponding to a recession speed of 137,000 km/s and a look-back time of about 4.5 billion years. The Berkeley group had demonstrated that supernovae might after all have something to offer cosmologists.

As an exercise in positive thinking, the group rebranded itself as the Supernova Cosmology Project (SCP). Pennypacker departed. Responsible budget management had not been numbered among his superpowers. Muller had stepped back, so (with the veteran Goldhaber's support) Perlmutter became the SCP leader. The project survived a funding review in the autumn of 1994. Although the CfPA cut its funding contribution, the gap was bridged with increased funding from the LBL.

* Supernovae are named with the prefix 'SN', the year of their discovery, and a suffix running from A to Z, then aa to az, ba to bz, ... za to zz. The plural of SN (supernova) is SNe (supernovae).

Perlmutter devised a new search strategy. Collect 100 images, each of hundreds of galaxies and possibly clusters of galaxies, just after a new moon. Wait 2½ weeks and collect a second set of images just before the next new moon. Subtract the second set from the first, image by image. Any residual spots left after subtraction are supernova candidates whose locations can be passed immediately to other astronomers who could then measure the spectra and light-curves using telescopes on which time had already been reserved. He called it the 'batch' method. Observations carried out in a single night could sample a statistically significant tens of thousands of galaxies, enough to guarantee catching at least one supernova early in its light-curve, while it is still brightening.

It worked. In 1994 the SCP found six Type Ia supernovae, SN 1994F, G, H, al, am, and an, with redshifts between 0.32 and 0.425. 'This epoch ends in my mind with a celebratory party at Gerson Goldhaber's [home] in the Berkeley hills, where we have a bottle of champagne for each of the half dozen SNe discovered in a batch'.[16]

High-*Z*

The astronomers were not overly impressed by the SCP's achievements. There was nothing particularly new in Perlmutter's search strategy. Kirshner, chairman of Harvard's astronomy department, a leading expert on supernovae, and a member of the Berkeley group's external advisory board, remained deeply sceptical. As far as he was concerned, the Berkeley physicists were still not asking the right questions. The issue was not whether high-*z* supernovae could be found, but whether they could serve as genuine standard candles.

The Berkeley group's paper on SN 1992bi had been sent to him for review. His initial delight turned to dismay, and he withheld approval (as did a second referee). The paper '... seemed to minimize three things. That photometry is hard. That SN [Type] Ia are not all alike. And what about dust?'[17] To account properly for the reddening effects of inter-stellar dust it was necessary to observe the light-curve through different coloured filters, which the Berkeley group had not done. The paper was eventually approved for publication by Sandage, with more modest claims, three years after its submission.[18]

Kirshner believed that the right way to proceed was by first char-acterizing nearby supernovae to build up an understanding of their light-curves. The fact that these did not appear standard didn't neces-sarily mean that they couldn't be standardized. Studies by astronomers at the Cerro Tololo Inter-American Observatory in northern Chile and the National Astronomical Observatory on Cerro Calán near Santiago found that the light-curves of brighter supernovae decline more slowly, those of dimmer supernovae decline more quickly. The *shape* of the light-curve is an indicator of brightness.

Adam Riess, a PhD student jointly supervised by Kirshner and William Press, an expert on numerical analysis, developed methods to determine both the intrinsic brightness of a supernova and ('a new idea, probably the first I had had on my own')[19] the reddening effects of dust. The resulting multi-colour light-curve shape analysis reduced the scat-ter in the measurements of the distance modulus of nearby supernovae from 40% to 15%. This was significant. The improvement reduced the number of high-z supernovae required to produce meaningful estimates of H_0 and the deceleration parameter by a factor of nearly 6. Kirshner felt vindicated.

In early March 1994, Perlmutter had tracked Kirshner down to the Mount Hopkins Observatory in Arizona. A breathless phone call ensued. Could Kirshner drop everything and measure the spectrum of SN 1994G? Perlmutter was persuasive, and relentless, and Kirshner, Riess, and Peter Challis from the CfA recorded 'the best spectrum ever taken of a high-redshift supernova'.[20] Riess later recalled that the experi-ence 'sparked an interest in all of us to fish for supernovae in the higher redshift waters'.[21]

When Brian Schmidt, another Kirshner PhD, and Nicholas Suntzeff, an astronomer at Cerro Tololo, suggested they form a rival supernova search team, the Harvard astronomers didn't hesitate. The High-Z Supernova Search Team included astronomers involved in the Calán/Tololo collaboration, the European Southern Observatory, and Harvard/CfA. Fourteen astronomers from six different institutions. The SCP now had some tough competition.

The High-Z team found its first supernova in April 1995, setting a new redshift record of $z = 0.478$. But at a conference in Spain that summer, at which both the rival teams were represented, it was clear that the SCP was benefitting from its six-year lead. The High-Z team had some catching up to do.

Return of the Cosmological Constant?

The Calán/Tololo collaboration submitted an estimate of $H_0 \sim 62 - 67$ km/s/Mpc to the *Astronomical Journal* in July 1994, and a revised version of the paper was received by the journal on 24 August.[22] Mario Hamuy at Cerro Tololo had agreed to share the data gathered by the collaboration with Riess at the CfA, on the understanding that the collaboration would publish their results first. Following their paper's acceptance, Hamuy gave Riess the green light and a few weeks later, Kirshner, Press, and Riess submitted their estimate of $H_0 = 67 \pm 7$ km/s/Mpc derived using the light-curve shape analysis.[23] But they submitted their paper instead to the *Astrophysical Journal Letters*. This was received on 6 September and accepted on 13 October. Although the Calán/Tololo paper had been accepted two months previously, and Hamuy had chased the editor of the *Astronomical Journal* to get their paper into print as quickly as possible, both papers were published in January 1995.

Priority was clearly established by the dates of receipt and acceptance by the respective journals, and Kirshner, Press, and Riess had thanked the collaboration for 'the opportunity to study their outstanding data before publication'. But the simultaneous publication of the two papers meant that they would likely always be cited together. Some in the Calán/Tololo collaboration felt bitter about the outcome, not because their priority had not been respected, but rather from what some perceived as a lack of sensitivity to the research ambitions of the collaboration.

Combined with the HST Key Project estimate of $H_0 = 80 \pm 17$ km/s/Mpc, it seemed that the direction of travel was towards a figure of about 70 km/s/Mpc, perched right in between the battle lines of 50 and 100 km/s/Mpc at the heart of the Hubble wars. In an Einstein–de Sitter universe, this implied an age of less than 10 billion years, younger than the universe's oldest stars. And a matter density parameter $\Omega_M \sim 0.2 - 0.3$ (including dark matter), much too small to be consistent with a universe driven flat by inflation, as seemingly confirmed by the COBE results.

A few cosmologists were becoming exasperated, to the point where they were ready to think the previously unthinkable. Lemaître had obviously been wrong about his primeval atom, but could he have been right about Einstein's cosmological constant? Admitting a cosmological constant or non-zero vacuum energy could potentially make all the problems go away. The expansion of the post Big Bang universe would indeed have decelerated, but only to a point where the vacuum energy of 'empty' space becomes dominant, driving an *accelerating* expansion thereafter. Recall that Lemaître had written in 1931: 'It is doubtless in this third period that we find ourselves today, and the acceleration of space which followed the period of slow expansion could well be responsible for the separation of the stars into extra-galactic nebulae'.[24] If $\Omega = \Omega_M + \Omega_\Lambda$, a vacuum energy density parameter of $\Omega_\Lambda \sim 0.7 - 0.8$ would ensure $\Omega = 1$, as desired. This also pushes the age of the universe from less than 10 billion years to a much more palatable 13.5 to 15 billion years.[25]

These unthinkable thoughts and assessments of the evidence to support them were published in October 1995 by Ostriker and Steinhardt,[26] and in November 1995 by Lawrence Krauss, then at Case Western Reserve University in Cleveland, Ohio, and Mike Turner.[27] Krauss and Turner provocatively titled their paper 'The Cosmological Constant is Back'.

The community was broadly unimpressed. 'These lines of evidence, while persuasive to a few investigators, left most of the community skeptical'.[28]

Hunting for Bumps

Schmidt and Suntzeff had founded the High-Z search team, and their names had headed the first observing proposal to Cerro Tololo. But the team needed a leader, and Suntzeff already had a full-time job at the observatory. Kirshner was an obvious choice, but it seems that Suntzeff tended to blame Kirshner for the bad feeling surrounding the simultaneous publication of estimates of H_0 based on the Calán/Tololo collaboration data. Suntzeff quietly lobbied for Schmidt, and Schmidt was duly selected. He had gained his Harvard PhD only three years previously, under Kirshner's supervision.

Both the High-Z search team and the SCP applied for observing time at Cerro Tololo, and were equitably awarded alternate nights. As an astronomer resident at the observatory, Suntzeff was now obliged to support the efforts of his rivals. These he found impressive, but he was dismayed by how far ahead the SCP team had got. By the autumn of 1995 they had gathered data on another 11 supernovae.

In January 1996, Perlmutter approached Robert Williams, the director of the STScI. He argued that the batch method encouraged a level of confidence in the detection of supernovae such that observing time on the HST could be reserved in advance to study them, with unprecedented resolution. Williams suggested he submit a proposal. Kirshner was firmly against the idea—the scientific case for the HST had been based on the notion of doing astronomy that could only be done from Earth's orbit, and both the SCP and High-Z teams had shown that observations of supernovae could be done perfectly well with ground-based telescopes. But Williams liked the proposal, and it quickly dawned on the High-Z team members involved in the discussion that they too might gain access to HST time. Once again this meant treating the rival teams equitably, which incensed Perlmutter (the SCP had been at this game for six years already and it had been his idea, after all). Williams allocated some of his discretionary time to both teams and agreed that the SCP could go first.

In the meantime, the SCP decided to push ahead with analysis of the data from their first seven high redshift supernovae, combined with the data on nearby supernovae from the Calán/Tololo collaboration. This appeared to support a cosmology with $\Omega_M \sim 0.94$ and $\Omega_\Lambda \sim 0.06$, both with substantial errors.[29] These were preliminary estimates, at best. The SCP had another 21 supernovae to include in their analysis, as well as data to be gathered using the HST. The team would soon be scrambling to revise its conclusions.

By September 1997 the SCP had gathered data on 38 high redshift supernovae. The data set included SN 1997ap, numbered among three supernovae studied using the HST, with another record-breaking redshift $z = 0.83$. The preliminary results were scattered, but proper error analysis would require several months of further observations. Goldhaber had spent 30 years hunting for 'bumps' in data distributions that hinted at the existence of new particles. In particle physics, the data referred to the particle mass (as E/c^2). Goldhaber figured he could use the same approach to hunt for a bump in the distribution of the matter density parameter, Ω_M.

In a flat universe, a plot of brightness (a proxy for distance) vs redshift has a slightly differently shaped 'Hubble curve' depending on the value of Ω_M. Goldhaber plotted Hubble curves for $\Omega_M = 0, 0.2, 0.4, 0.6, 0.8,$ and 1.0, shown in Fig. 31(a). He then plotted the supernova data on top of these.[30] The result does not look encouraging.

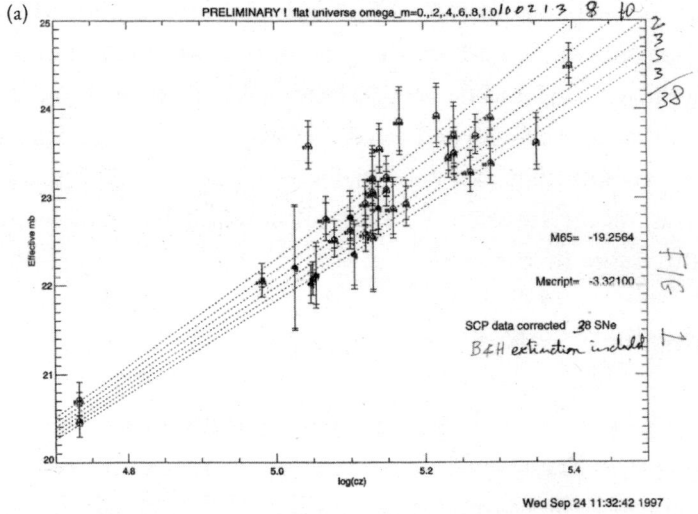

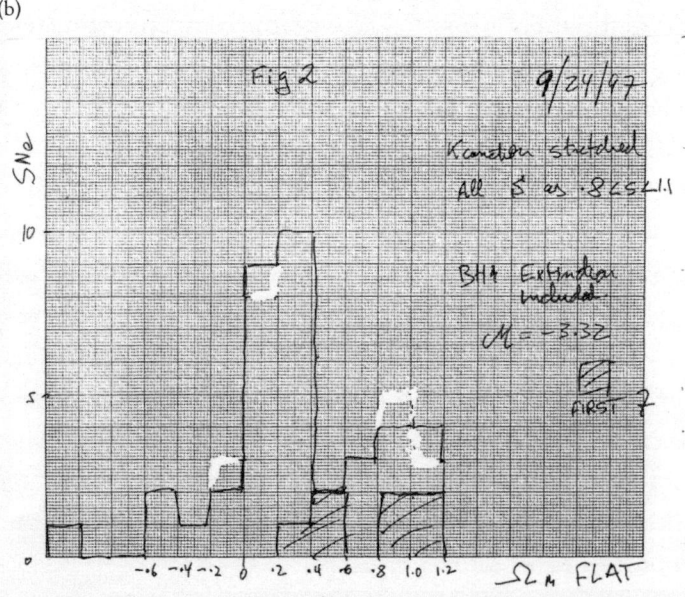

Fig. 31 Goldhaber compared the SCP data for 38 supernovae with Hubble curves calculated for $\Omega_M = 0, 0.2, 0.4, 0.6, 0.8$ and 1.0, (a). Although the scatter in the data appears unpromising, he simply counted the number of data points falling in the range $\Omega_M = 0 - 0.2, 0.2 - 0.4$, etc., and plotted the results in a histogram, (b). The data appeared to be pointing firmly towards the result $\Omega_M \sim 0.2 - 0.3$.

The scatter in the data appears too large to draw any meaningful conclusions. Some of the data points even imply an impossibly negative value for Ω_M.

But Goldhaber now simply counted (by hand) the number of data points that fell in the intervals $\Omega_M = 0 - 0.2, 0.2 - 0.4$, etc. and plotted the results in a histogram, Fig. 31(b). Two things leapt out. Of the data from the team's first seven supernovae, shown as the cross-hatched blocks in Fig. 31(b), several appeared to be high-Ω_M outliers. The peak in the histogram, with 10 data points, favoured a lower value of Ω_M between 0.2 and 0.4, and 9 data points favoured $\Omega_M \sim 0 - 0.2$. Taken together, these imply $\Omega_M \sim 0.2 - 0.3$. He shared these graphs at an SCP team meeting on 24 September.

This estimate for Ω_M was entirely consistent with previous estimates. But the supernova data were capable of telling more of the story. The Hubble curves in Fig. 31(a) were calculated for a flat universe with no assumptions about the value of Ω_Λ. But equivalent curves could be readily calculated for a universe with $\Omega_\Lambda = 0$. Recall that the deceleration parameter q_0 for such a universe is simply $1/2\Omega_M$. When mapped onto these curves, of the 38 data points, 25 indicated a strikingly unphysical *negative* matter density parameter, of the order $\Omega_M \sim -0.4$. The only way to make any sense of this was to adopt a *different cosmology* with a positive value for Ω_Λ. This would mean that the deceleration parameter is now given by $1/2\Omega_M - \Omega_\Lambda$, reflecting the tension between the gravitational pull of matter and the 'anti-gravitational' push of vacuum energy. In a flat universe, the value of $\Omega_M \sim 0.2 - 0.3$ indicated by Fig. 31(b) suggested that Ω_Λ take a value between 0.7 and 0.8, implying q_0 has values between -0.55 and -0.7.

A negative deceleration is actually an *acceleration*. Perlmutter and Goldhaber separately presented the SCP results, updated to include two more supernovae, at colloquia and seminars delivered to the physics departments in San Diego, Berkeley, and Santa Cruz, and at the Institute for Theoretical Physics at Santa Barbara, in October and December 1997.

Although these were not public announcements, evidence suggesting a positive cosmological constant and density parameter $\Omega_\Lambda \sim 0.7$ might have been expected to send shock waves through the astronomy and cosmology communities, and would most surely have been reported in the mainstream media. However, the SCP data were presented as preliminary, subject to further revision, and were not yet published. The team had only recently published other preliminary results suggesting $\Omega_M \sim 0.94$ and $\Omega_\Lambda \sim 0.06$. The presentations were conservative, cautious, and qualified. At Berkeley, Perlmutter explained that their data held the promise of 'some rather striking consequences for physics'. Sitting in the audience for Perlmutter's talk at Santa Cruz was American physicist Joel Primack, who had worked on theories of cold dark matter in the 1980s. Primack acknowledged that the results were 'earthshaking ... if true'.[31] For now, it seems, judgement was suspended.

Dark Energy

Riess had completed his PhD at Harvard in 1996, and had cast around for a post-doctoral position. There was no funding available for a post-doc with the High-Z team, and for a time he considered an offer to join the SCP. But then he was offered a prestigious Miller Fellowship at the University of California at Berkeley. This allowed him the freedom and flexibility to continue working with the High-Z team. He was also able to join astronomer Alexei Filippenko, who had recently defected from the SCP to High-Z, bringing with him access to the Keck Observatory at the summit of Mauna Kea in Hawaii.

By November 1997 the High-Z team had recorded fewer supernovae than the SCP, but the team members believed that what they lacked in quantity was more than compensated for by the quality of their data. Their data set included four supernovae observed using the HST, with SN 1998I recording $z = 0.89$. Riess carried out a preliminary 'first-pass' analysis. What he found was greatly puzzling. He examined the data in the context of a universe with $\Omega_\Lambda = 0$: '... what I initially measured and wrote in my lab notebook in the fall of 1997 was stunning! The only way to match the expansion rate I was seeing was to allow the universe to have a "negative" mass'.[32] He estimated $\Omega_M = -0.36 \pm 0.18$, which made no sense. Unless, of course, Ω_Λ is greater than zero and the universe is actually accelerating.

Both teams were unaware of what the other had found. Both teams now moved quickly to check and re-check their data and analysis. Restoring Einstein's cosmological term to the equations of the universe was a bold move, which could either make careers (and win Nobel prizes) if it was right, or wreck careers if it proved to be wrong. On 8 January 1998 Perlmutter presented the SCP results—now fully analysed and incorporating detailed error corrections—at a meeting of the American Astronomical Society in Washington, DC. But the press release accompanying Perlmutter's lecture speaks of the 'fate of the universe' and mentions the cosmological constant only in passing.[33] A report of the lecture published in the *New York Times* follows the line provided by the press release.[34] A report published in *Science* magazine was more forthright about the cosmological constant, 'If they [the results] hold up'.[35] Although it logically follows from $\Omega_\Lambda \sim 0.7$, the magic words 'accelerating universe' are conspicuous by their absence.

Earlier that day Schmidt had sent Riess an email from Australia agreeing the details of their calculations and with the subject line 'Hello Lambda'.[36] Riess shared his findings with the rest of the High-Z team and there followed a flurry of email exchanges arguing variously for caution or boldness. Kirshner didn't like it at all. Suntzeff urged Riess (who was at this time on his honeymoon): '... my god, get it out! ... I mean this seriously—you probably never will have another scientific result that is more exciting come your way in your lifetime'.[37] Schmidt argued that it was wrong not to publish a result just because they didn't like it.

Goldhaber and Perlmutter presented the SCP's evidence for the accelerating universe—now based on observations of 42 Type Ia supernovae—at a conference on dark matter held in Marina del Rey in

California on 18 February 1998. These were followed by a presentation from Filippenko based on the High-Z results for 16 supernovae. The dam now burst. 'Wary astronomers ponder an accelerating universe', ran the *New York Times* headline on 3 March. Perlmutter acknowledged his reticence at the January meeting: 'We were trying to be very conservative until we had more observations', he explained.[38]

The teams published papers in 1998 and early 1999.[39] 'Our teams, especially in the US, were known for sort of squabbling a bit', Schmidt explained at a press conference some years later. 'The accelerating universe was the first thing that our teams ever agreed on'.[40]

Arguments based on the formation and stability of galaxies, or large-scale structure derived from redshift surveys, were not easily waved away. But the supernova data provided a much more direct measure, based on observations of distant objects with long look-back times. Sandage had insisted that understanding the evolutionary history and ultimate fate of the universe—or choosing between different cosmological models—was all about measuring two parameters, H_0 and q_0. Here—finally—was a direct measure of q_0. The existence or otherwise of a cosmological constant determined the brightness of the high redshift supernovae being observed, and the supernovae were *dimmer*—their host galaxies further away—than could be accounted for with a decelerating universe consisting of matter alone.

The conclusion was sobering. Yes, the return of the cosmological term fixed more than a few of the problems that had nagged at inflationary Big Bang cosmology. But to the unfathomable substance that was dark matter astronomers had now added a deeply unfathomable vacuum energy. At 70% of the matter–energy density this was by far the largest component of the universe.

Could the cosmological vacuum energy be simply equated with the energy associated with the kinds of quantum fluctuations characteristic of the Casimir effect? Physicists tried. But attempts to calculate the energy density of the vacuum arising purely from quantum fluctuations produced a result that is a hundred billion billion googol (10^{120}) too large. That's got to be the worst theoretical prediction in the entire history of science. The problem remains unresolved.

To avoid confusion with the vacuum energy implied by quantum fluctuations, Turner scratched around for an alternative name. 'Funny energy' wasn't serious enough. He settled on the rather unimaginative name 'dark energy', which stuck. It would seem that what we considered as 'the universe' not so very long ago—a universe of stars and galaxies (baryonic matter)—accounts for only 4% of the universe that was now revealed to astronomers.

A few years later Riess and his colleagues unearthed SN 1997ff, discovered in the Hubble 'deep field', a small region in the constellation Ursa Major containing over 3000 objects, almost all galaxies. It had been serendipitously photographed during the commissioning of a sensitive infrared camera. Its redshift of $z = 1.7$ set another new record, with a look-back time of 9.7 billion years. Whereas older and older supernovae had appeared fainter and fainter than simple extrapolation would suggest, SN 1997ff bucked this trend and appeared a little *brighter*. It was the first glimpse of a supernova that had been triggered in a period when the expansion rate of the universe had still been decelerating.[41]

The result suggested that about five billion years ago, the expansion rate 'flipped' (see the upper dashed curve in Fig. 32). As expected, gravity had slowed the rate of expansion of the post-big bang universe until it reached an inflection point at which it had begun to accelerate, just as Lemaître had proposed in 1931. Contrast this with the Hubble curve for an Einstein–de Sitter universe with $\Omega_M = 1$ and $\Omega_\Lambda = 0$, which slowly decelerates to infinity.

The gamble, such as it was, paid off. Perlmutter was awarded a half-share of the 2011 Nobel prize for physics, with Schmidt and Riess sharing the other half between them 'for the discovery of the accelerating expansion of the universe through observations of distant supernovae'.

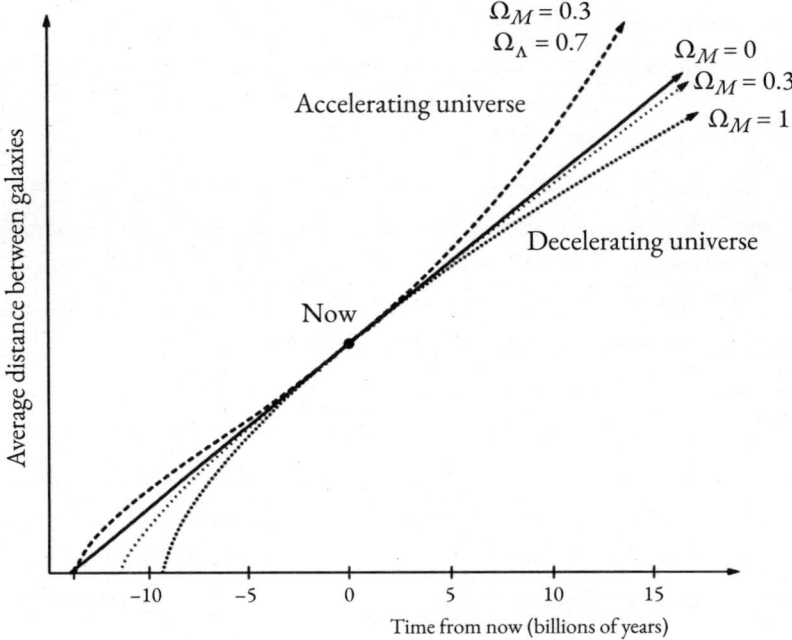

Fig. 32 Studies of very high redshift supernovae revealed more details about the history of the universe. Look-back times approaching 10 billion years revealed a moment when the expansion of the post-big bang universe (with $\Omega_M = 0.3$ and $\Omega_\Lambda = 0.7$, the upper curve in this figure) was still slowing.

The HST Key Project

The HST Key Project team used a variety of methods to determine H_0. The Cepheid distance scale was used to determine distances of nearby spiral galaxies up to 30 Mpc. This scale was then used to calibrate five independent secondary methods, including observations of Type Ia supernovae, and the empirical relation between galaxy luminosity and rotation speed deduced by Tully and Fisher. The results from each method were in good agreement and, when combined, yielded a value for H_0 of 72 ± 8 km/s/Mpc, a precision of 11%. Assuming $\Omega_M \sim 0.3$

Photo 17 In 2006, Saul Perlmutter (left), Adam Riess (centre), and Brian Schmidt (right) received the Shaw Prize in Astronomy. They would go on to receive the Nobel physics prize in 2011 'for the discovery of the accelerating expansion of the universe through observations of distant supernovae'.

and $\Omega_\Lambda \sim 0.7$ yielded an estimate of the age of the universe of 13 ± 1 billion years. The team published their final results in early 2001.[42]

They had beaten down the errors.

9

Concordance

Although the SCP and High-Z teams had rightly fretted about the potential repercussions of announcing evidence for an accelerating universe, the community embraced the result really rather quickly. It helped that two rival research teams had independently arrived at the same conclusion, much as experimental high-energy particle physics had long before adopted the strategy of establishing independent teams to study the data flowing from the same particle collider, using different detection techniques. For sure, there were still many mysteries, but it seemed that the astronomers and cosmologists were converging on the answer to the questions of what kind of universe we live in, and what its ultimate fate is likely to be.

The success of the supernova search teams had demonstrated that understanding the present (and predicting the future) depended critically on detailed studies of the past, aided by the look-back effect. The further back we could look, the better our understanding of the evolutionary history of the universe. But, of course, the ultimate source of historical information remained the cosmic background, with $z = 1110$ and a look-back time of 13.1 billion years, almost to the very beginning. Theoreticians insisted that the cosmic background had yet more secrets to give up. Unlocking these would require even more detailed study of its temperature anisotropies.

There are other sources of anisotropy aside from the quantum fluctuations revealed by COBE and the redshift and time dilation associated with the Sachs–Wolfe effect. If the primordial universe was a radiation-dominated plasma, it was expected to have released radiation that is moving relative to the overall expansion of the universe, and so there is a contribution to the temperature fluctuations arising from a Doppler effect. When written in its modern form, these physical processes occurring in the plasma just prior to recombination—the point of 'last scattering'—account for the first three terms in what is known as the *Sachs–Wolfe equation*. The fourth term relates to effects occurring both soon after recombination and many years later. Such effects may distort the pattern of temperature fluctuations along the 'line of sight' towards the instrument used to detect the radiation, and when added together are collectively known as the integrated Sachs–Wolfe effect.

But there is a further contribution overlaid on top of these that is in itself not only quite remarkable, but also offers insights on the physical composition of the universe. The theorists had an extraordinary tale to tell.

The Signature of the Universe

This was all about the physics of the universe as a high-temperature plasma, before recombination. The dark matter (whatever it is) and baryonic matter (protons and helium nuclei) pulled towards an over-density was surrounded by dense radiation that resisted compression. As the matter was drawn in, the radiation interacting with the baryonic matter and free electrons was drawn in with it. The radiation pressure built up, and the resulting competition between gravity and radiation pressure triggered acoustic oscillations—*sound waves*—in the plasma. These oscillations occurred wherever there were over-densities, triggered at different times in the early universe depending on how quickly matter accumulated. Imagine tossing a handful of pebbles into a still pond, such that they don't all break the surface at the same time, and watching as the ripples expand, first here and then there, before overlapping and merging. Now stretch your imagination to picture the scene in three dimensions, in which three-dimensional ripples appear at different times and places.

Alas, even if there had been someone around who could listen, these were not sound waves that could have been heard. Human perception has evolved to hear acoustic oscillations with frequencies between 20 hertz (cycles per second) and 20,000 hertz, propagating in air with a speed of around 340 m/s. This range of frequencies corresponds to wavelengths ranging from 17 metres to 17 millimetres. But the sound waves propagating in the primordial fluid moved at speeds of more than half the speed of light, with (as we'll see) wavelengths measured in millions of light-years.

Nevertheless, I still like to think of this as a period when the universe was *singing*.

The pressure wave developing in the plasma was manifested as a compression of the radiation-dominated fluid which expanded outwards. The negatively charged electrons were pulled along for the ride, dragging the positively charged baryons behind them. Because dark matter does not interact with radiation, it got left behind. The end result was a spherical wave of over-dense baryonic matter that expanded outwards, leaving a 'rarefaction'—a region of low matter density—in its wake. Sound waves produced early derived from smaller over-densities and were therefore of smaller amplitude and higher frequency (shorter wavelengths). They were heavily 'damped' as the radiation diffused away in a process first identified in 1968 by Joe Silk, and which is called 'Silk damping'. This meant that the wave was not sustainable much beyond a single compression–rarefaction cycle. Ultra-high-frequency sound waves travelling in air are unsustainable for similar reasons.

While all this was going on the universe was continuing to expand, and when it had cooled sufficiently to allow recombination the electrons were captured by the baryons, Thomson scattering ceased, and the radiation was released to form the cosmic background. This process is illustrated in Fig. 33, which shows a computer simulation of the motions of baryons and photons starting on the left with an over-density which triggers a compression wave. Initially, the baryons and photons moved outwards together, and the wave continued to expand for 100,000 years or so.[1]

Baryons

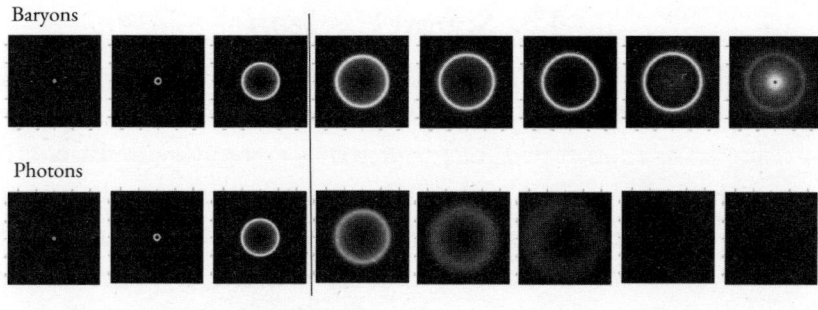

Photons

Recombination

Fig. 33 This computer simulation shows how a compression wave triggered by the build-up of radiation pressure expands outwards from an over-density. Initially, the baryons and photons move together, but recombination disconnects the photons which diffuse away, leaving a spherical shell of excess baryonic matter.

Recombination disconnected the photons and these diffused away. The radiation pressure and the speed of sound fell dramatically, leaving a spherical shell of baryonic matter 'frozen' in place, like a line of flotsam carried up a beach by a high tide (but, again, in three dimensions). Over time, the dark matter left behind at the initial over-density accumulated more baryonic matter and the shell of excess baryonic matter accumulated more dark matter.

The early, heavily damped waves would have left a small-scale imprint on the distribution of matter. But later waves that built just before recombination were predicted to be of larger amplitude and lower frequency (longer wavelength). Regions of high matter density associated with a compression wave would have produced hotter background radiation—a positive fluctuation in temperature. Regions of low matter density associated with a rarefaction would have produced cooler background radiation—a negative fluctuation in temperature. So, frozen into the cosmic background radiation is an imprint of the distribution of matter just a few hundred thousand years after the Big Bang. This is the 'signature of the universe'.

The Sound Horizon

What kind of distance scale are we talking about? This is best answered by reference to another perspective offered by computer simulations. In Fig. 34 the 'mass profiles' of the different constituents of the universe are plotted against distance measured in Mpc for six snapshots in time.[2] Think of these plots as a kind of 'time-lapse' movie, and note that the radius plotted along the x-axis in Fig. 34 is expressed in so-called 'co-moving' coordinates, such that the ongoing expansion of the universe is 'built-in' to the distances in each snapshot.

At early times, just 14,433 years after the Big Bang, the compression wave was underway, pushing baryons and photons outwards from the over-density and leaving the dark matter behind—Fig. 34(a). Neutrinos are notorious for not interacting with anything, so they simply diffused away, producing a 'cosmic neutrino background'. Neutrinos are thought to have decoupled from matter when the universe was just one second old. Because it has existed longer and has been subjected to more of the universe's expansion history, the cosmic neutrino background is cooler than the cosmic microwave background, with a temperature of about 1.95 K. The PTOLEMY project, begun in 2019, aims to provide a scalable design for a cosmic neutrino background detector.

After 230,000 years (Fig. 34(b)) the baryons and photons were pushed out beyond 100 Mpc through a combination of radiation pressure and the overall expansion of the universe. Recombination occurred at about 380,000 years, so after 570,000 years (Fig. 34(c)) the disconnected photons were starting to diffuse away, forming the cosmic background radiation and leaving an excess of baryons in a spherical shell with a radius of about 150 Mpc as measured from the initial over-density. This is called the *sound horizon*. The cosmic background radiation would then be expected to carry an imprint of this matter distribution. An excess of matter would give a slightly higher temperature. A dearth of matter would give a slightly lower temperature.

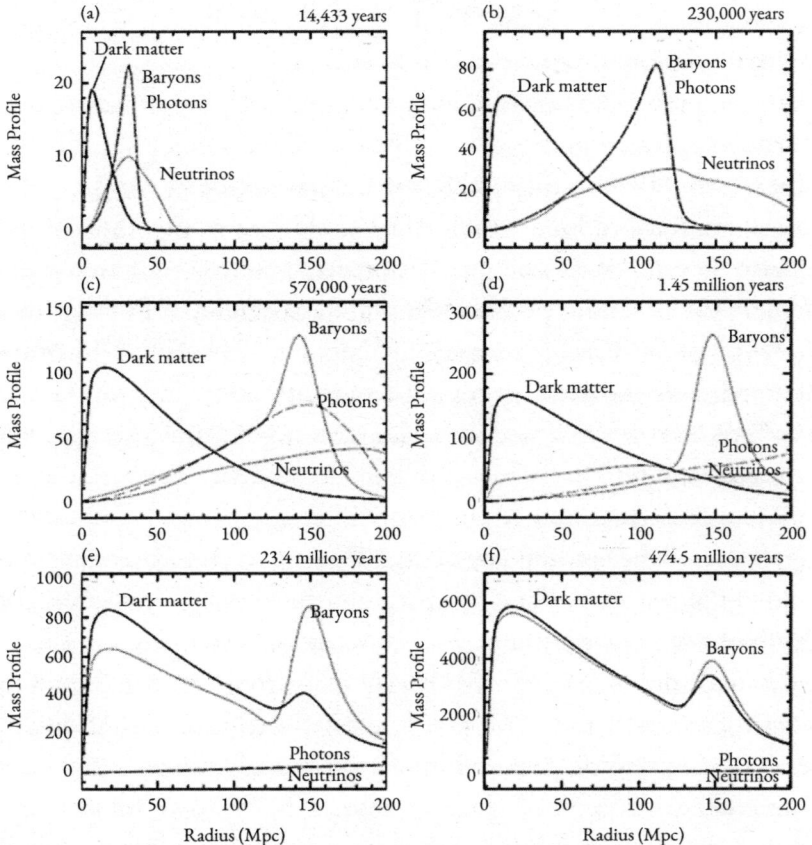

Fig. 34 Computer simulations of the mass profiles of the constituents of the early universe provide a kind of time-lapse movie of the effects of the compression wave as it expands outwards from the over-density. Note how the baryons and photons move outwards together in (a) and (b). Recombination occurs between (b) and (c), so in (c) the photons have disconnected and are diffusing away, leaving an excess of baryonic matter in a spherical shell with a radius of 150 Mpc, known as the *sound horizon*.

After 1.45 million years (Fig. 34(d)) the photons and neutrinos were disappearing from the picture, leaving prominent over-densities of dark matter and of baryonic matter positioned at the sound horizon. After 23.4 million years (Fig. 34(e)) the gravity of the dark matter had started to pull more dark matter and baryonic matter back towards it, and the excess baryonic matter in the shell had started to do the same. After 474.5 million years (Fig. 34(f)) the equalization of the dark matter and baryonic matter was almost complete, leaving a large over-density close to the centre and a secondary over-density at the sound horizon. This is about the time we would expect the first stars and galaxies to be forming.

The theorists realized that, if they could indeed be observed, patterns in the temperature fluctuations attributable to the acoustic oscillations would be sensitively dependent on the cosmology of our universe. Soviet physicists Rashid Sunayev and Yakov Zeldovich at the Institute of Applied Mathematics in Moscow were first to publish detailed calculations in 1970, followed shortly (and independently) by Peebles at Princeton and Jer Yu, Peebles' first graduate student.[3]

When translated into temperature differences, at its maximum the amplitude of the last and largest compression wave is governed by the ratio of the density of baryonic matter to the density of radiation. The depth of the associated rarefaction is attenuated compared to this, by an amount determined by the density of baryonic matter. The wavelength of this last sound wave, as reflected in the difference in the distance scales of compression peak and rarefaction trough, is sensitively dependent on the curvature of space. And, because what we see in the sky today is the result of about 13 billion years of expansion since recombination, the value of the Hubble constant is also firmly embedded in the description.

The theorists understood that analysis of the impressions left behind by the acoustic oscillations could potentially tell us precisely what kind of universe we live in: its total density, flatness, the nature (cold, warm, or hot) and density of dark matter, the value of Λ, the density of baryonic matter, and the Hubble constant, H_0.

If this scenario provided a broadly correct description of the early universe, it implied that baryonic matter carried out to the sound horizon would leave a concentrated sphere of matter about 150 Mpc or so distant from the centre. This would be where filaments or 'walls' of galaxies would eventually form, shaping the large-scale structure of the universe. It is surely no coincidence that the nearest objects in the Great Wall are found about 100 Mpc, and the furthest objects 170 Mpc, distant. Here was an explanation for the patterns of walls and voids in the distribution of galaxies that would become known as the *cosmic web*.

The TT Power Spectrum

Look all you like, but you won't see these patterns in the all-sky map of temperature differences in the cosmic background radiation as revealed by COBE (Fig. 29). To see them requires some considerable mathematical manipulation of the data. As these manipulations are important to understanding how cosmological parameters, including H_0, are derived from the analysis of the temperature data, it's worth taking a little time to appreciate what's involved.

To understand how we might be able to 'see' the acoustic oscillations, it helps to remember what we're looking for. If the theorists have it right, then the matter over-densities (compression) and under-densities (rarefaction) left an impression in the pattern of the distribution of matter and hence temperature fluctuations in the cosmic background radiation, on a distance scale of the sound horizon—about 150 Mpc—and smaller. These temperature fluctuations (let's call them ΔT) can be positive ($+$), signalling a slightly hotter region of high matter density compared with the average, or negative ($-$), signalling a slightly cooler region of low matter density compared with the average.

Suppose we look in two different directions in the sky, m and n, as measured from the instrument on board a high-altitude balloon, U-2 'spy plane', or satellite. The angle between these directions is θ (Greek theta, see Fig. 35). We can see immediately from this that the size of the angle between the two directions (the *angular scale*) defines a length or distance scale in the sky. We measure the size of the temperature

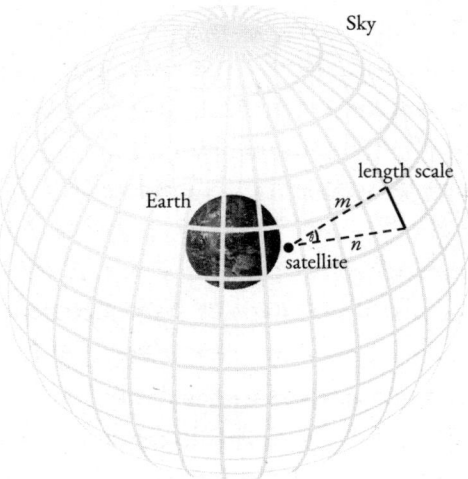

Fig. 35 Looking in two different directions, m and n, separated by an angle θ measured from the instrument (here depicted aboard a satellite), defines a length scale in the sky. When averaged over all possible directions with the same angle θ across the sky, the product of the temperature differences $\Delta T_m \Delta T_n$ defines a two-point correlation function.

fluctuations in these different directions, ΔT_m and ΔT_n, and multiply them together. We repeat this for all possible directions across the sky at the same angle, and average all the contributions. This gives us a *two-point correlation function*. We then repeat this for all possible angles spanning 360°.

To see how this works it helps to imagine what we would get if the temperature fluctuations happened to be completely randomly distributed across the sky, both in terms of absolute size and sign. When the temperature differences are multiplied together as $\Delta T_m \times \Delta T_n$, for a given angle θ we would expect equally as many instances of positive–positive $(++)$, positive–negative $(+-)$, negative–positive $(-+)$, and negative–negative $(--)$ contributions. Any positive number multiplied by a negative number gives a negative result, so the $+-$ and $-+$ contributions are overall negative. These will cancel all the positive contributions from the equally probable instances of $++$ and $--$, such that the overall correlation is zero: a completely random distribution gives zero correlation.

We can see from this that any small excess of $++$ and $--$ contributions, which we might expect from regions in the sky where the baryonic matter was compressed and rarefied, will give rise to an overall positive correlation. This will happen for angular scales which correspond to the length scales of the acoustic oscillations, and especially the length scale of the sound horizon.

We're dealing here with patterns on the surface of a sphere, and in a standard approach to the analysis of such patterns, we express the two-point correlation function as a sum of so-called Legendre polynomials, named for French mathematician Adrien-Marie Legendre, who discovered them in 1782. A polynomial is a mathematical function constructed from variables (such as the eponymous x, or in this case the angle θ) and numerical coefficients. Terms in each function are added or subtracted, and may involve integral powers of the variables (x^2, x^3, or $\cos^2\theta$, ...). The Legendre polynomials form a set of increasing complexity, each governed by a running index of integer numbers or 'orders', l, such that $\theta \sim 180°/l$. The results for different orders are called *spherical harmonics* and, because the correlation is derived from the product of temperature differences (typically measured in millionths of a degree) in the different directions, a chart of this correlation vs l is called the *TT power spectrum*.

In April 1992 Smoot had presented data related to the order $l = 0$, the spherical all-sky map of the background which is dominated by microwave and infrared emission from the planc of the Milky Way galaxy. He went on to present results for order $l = 1$ ($\theta \sim 180°$), the dipole anisotropy first observed aboard the U-2 'spy plane' missions. Both of these are typically subtracted from the all-sky map in order to reveal the underlying pattern of anisotropies, as shown in Fig. 29. The audience was anticipating data for order $l = 2$ ($\theta \sim 90°$), a quadrupole pattern, but were stunned when he presented data for further orders up to $l = 20$, referred to collectively as 'multipoles'. The $l = 20$ pattern, corresponding to $\theta \sim 9°$, approached the limit of the angular resolution of the differential microwave radiometer installed aboard the COBE satellite.

In the TT power spectrum, the horizontal axis or abscissa is the increasing order l (sometimes $\log l$), which means that the spectrum shows correlations on *decreasing* angular scales, and hence decreasing length scales. The longest-wavelength acoustic oscillation, which also happens to be the last before recombination and is depicted in the computer simulations shown in Figs. 33 and 34, is therefore expected to appear at lower orders of l than shorter-wavelength oscillations which occurred earlier in the universe's history. This was at least encouraging: the most prominent acoustic oscillation should be seen at larger angular scales. But estimates quickly showed the nature of the challenge.

Translating the length scale defined by the sound horizon into an angular scale and an order l demands the assumption of a specific cosmology. The relation depends on the density of matter (both baryonic and dark), Ω_M, and the Hubble constant.[4] If we assume a flat universe with $\Omega_M \sim 0.3$, and a Hubble constant of 72 km/s/Mpc, then a sound horizon of about 150 Mpc (Fig. 34) implies an angular scale $\theta \sim 0.6°$. This is more than 10 times the angular resolution that had been available from the differential microwave radiometer aboard COBE.

And, by the way, good luck spotting by eye a slight excess correlation in positive and negative temperature fluctuations in the all-sky map on angular scales of less than 1 degree.

BOOMERANG

But the challenge had been set. COBE was shut down towards the end of 1993, having fulfilled its mission objectives. Two years later, a team of astrophysicists led by Charles Bennett at NASA's Goddard Space Flight Center submitted a competitive proposal for a new satellite mission. Bennett had served as the Deputy Principal Investigator of the DMR experiment aboard COBE. The objective of the proposed Microwave Anisotropy Probe (MAP) mission was to study the temperature fluctuations of the cosmic background radiation with sub-degree angular resolution.[5] The team also included John Mather, Gary Hinshaw, and David Skillman from NASA, and Norman Jarosik, Lyman Page, David Spergel, and David Wilkinson from Princeton, with contributions from Stephan Meyer at the University of Chicago, and Ned Wright at the University of California Los Angeles. In April 1996 NASA approved the proposal for a further 'definition study' as a potential mid-class *Explorer* mission.

The hunt for the acoustic oscillations in the TT power spectrum was now well and truly on, and some astronomers had no desire to wait for MAP to launch. You might be tempted to think that there was little to be done, as all the answers must surely have had to await another, more sensitive, space-based instrument. But there were ways to go beyond the angular resolution afforded by COBE using ground-based and balloon-borne instruments, admittedly with shorter observing times and covering more limited areas of the sky. All it took was a little ingenuity.

In 2000, Eric Gawiser at the University of California at Berkeley and Joe Silk, now at Oxford University in England, published a review summarizing the results of as many as 75 different experimental studies of the anisotropy—including COBE—that had been reported since 1992. The first acoustic peak in the TT power spectrum was now clearly visible. I don't propose to account for all of these results here, but will instead use the example of the Balloon Observations Of Millimetric Extragalactic Radiation ANd Geophysics (BOOMERANG) experiment to convey something of the sense of excitement that built as the acoustic peaks were gradually revealed.

BOOMERANG was an Italian–American collaboration, with principal investigators from the University of Rome La Sapienza and Caltech. It was proposed in 1995[6] and launched by NASA's National Scientific Balloon Facility on 29 December 1998 from Williams Field, at the McMurdo research station on the shore of McMurdo Sound in Antarctica. The balloon rose to its floating altitude of 120,000 feet (about 37 km) and observations began about three hours later. Over the following 10 and a half days the balloon completed an anticlockwise circuit above the South Pole. It then returned to the ground about 50 km from McMurdo station, from where it was recovered by helicopter.

The balloon carried a microwave telescope with an angular resolution capability of $\theta \sim 4°$ to $\theta \sim 0.3°$, the latter corresponding to an order $l \sim 600$. Measurements were made at four different microwave frequencies. Data taken at 150 GHz, covering just 1% of the sky, were analysed and reported in *Nature* magazine in April 2000. The first acoustic peak was now clearly visible, at an order $l = 197 \pm 6$—see Fig. 36(a)—corresponding to an angular scale of $\theta \sim 0.9°$.[7] The data were insufficient to define all the cosmological parameters, but provided enough detail to conclude that cold dark matter cosmologies in which the densities of matter and dark energy add up to 1 (i.e., $\Omega = \Omega_M + \Omega_\Lambda \sim 1$) afford the better description.

The universe is definitely flat.

Andrew Lange, the US team leader of the BOOMERANG experiment, explained it like this: 'Only five years ago we were arguing whether the density of the universe is 0.3 or 1. For a flat universe you need a density of 1. Now Boomerang tells us that the density is in fact very close to 1. And if the geometry is confirmed to be flat, that will be strong support in favour of inflation. The enormous expansion at the very beginning would have stretched the geometry of space until it was perfectly flat'.[8]

The peak shown in Fig. 36(a) is evidence for the last and longest-wavelength compression wave, but the data provide insufficient evidence for the second feature in the TT power spectrum, corresponding to the smaller-scale rarefaction. Over the next few years the larger data set from BOOMERANG was analysed, and the angular resolution was further extended to $\theta \sim 0.2°$. Data from four different detectors each

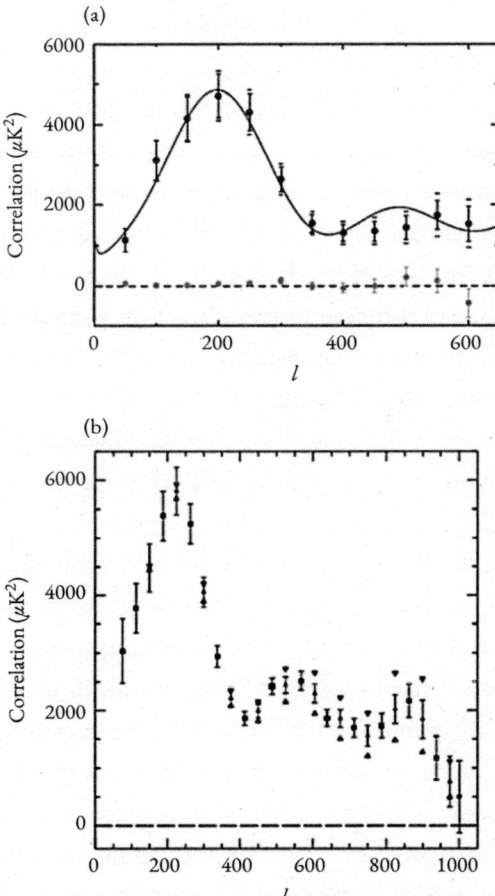

Fig. 36 Early analysis of the data from the BOOMERANG experiment reported in 2000 provided evidence for the first acoustic peak in the TT power spectrum, (a). Subsequent analysis of a larger data set reported two years later provided clear evidence for further acoustic oscillations at larger order (smaller angular scales), (b).

sampling about 1.8% of the sky were combined. The results, shown in Fig. 36(b) now clearly showed both the rarefaction and the next compression wave, corresponding to an earlier oscillation with a shorter wavelength.[9]

Concordance

But the scientists were now confronted with another analysis problem. When fitting a straight line using the equation $y = mx + c$, we're actually fitting two parameters, the slope of the line m and the constant c, to a series of data points, x, y. The TT power spectrum is a complex curve, and its theoretical description is based on a much larger set of parameters, including the total density parameter Ω, the products of the baryon density and dark matter density and the square of the dimensionless Hubble constant, $\Omega_B h^2$ and $\Omega_{DM} h^2$ (where $\Omega_M = \Omega_B + \Omega_{DM}$),* and the dark (or 'vacuum') energy density parameter Ω_Λ. Because these parameters are all inter-related, it's a little more difficult to produce a unique set of results from fitting the power spectrum alone.

One approach is to pin down (or 'constrain') the cosmological parameters as far as possible using data from other sources. In 2000, Max Tegmark, then at the University of Pennsylvania, Matias Zaldarriaga at the Institute for Advanced Study in Princeton, and Andrew Hamilton at the University of Colorado posted a couple of papers in which they chose to constrain a model consisting of 11 parameters using data from studies of the cosmic background radiation and from galaxy redshift surveys.[10]

IRAS, launched in January 1983, was a joint US (NASA), Dutch, and British mission. It was the first space-based observatory to conduct an all-sky survey at infrared wavelengths, which lasted 10 months. The IRAS Point Source Catalog Redshift (PSCz) Survey was published in 2000 and at this time listed redshift data for 18,351 galaxies. By fitting both the TT power spectrum derived from a variety of experiments (including COBE and the early BOOMERANG results) *and* galaxy redshift data *simultaneously* using the same 11-parameter model, Tegmark, Zaldarriaga, and Hamilton found that they could considerably improve the determination of the parameters themselves. They obtained $\Omega_B h^2 = 0.02$, $\Omega_{DM} h^2 = 0.13$, $\Omega_\Lambda = 0.62$, and $h = 0.63$. The latter value for h implies $\Omega_B = 0.05$, and $\Omega_{DM} = 0.33$. The total density parameter $\Omega = \Omega_B + \Omega_{DM} + \Omega_\Lambda$ adds up to 1.

---

* Remember, $h = H_0/(100 \text{ km/s/Mpc})$, so if $H_0 = 72 \text{ km/s/Mpc}$, $h = 0.72$.

They called it the 'concordance' model.

Although not every cosmologist was enamoured of the name, this kind of approach was broadly adopted for many subsequent analyses. Instead of simultaneously fitting the data from different sources, parameters derived from these sources are used as 'priors' to constrain the fitting of the TT power spectrum. The terminology is important. Values of the parameters gathered from other sources are not fixed in the analysis of the power spectrum. Rather, they are 'dropped in' to the analysis as prior estimates and allowed to find the values that provide the best fit to the power spectrum. What comes out of the analysis may be slightly different to what went in, but the idea is that what comes out will be both more accurate and more precise.

In their analysis of the acoustic peaks in Fig. 36(b), the BOOMERANG team examined the effect of increasing the constraints using priors from different sources, such as large-scale structure, redshift data from studies of Type Ia supernovae, a combination of both of these, fixing the Hubble constant at a value $h = 0.71 \pm 0.08$, constraining the universe to be flat ($\Omega = 1$), and further combinations of these. They found that assuming a range of values for h between 0.45 to 0.90 tightly constrains the value of Ω to be very close to 1. Constraining the analysis using large-scale structure and Type Ia supernovae data produced an estimate for h of 0.67 ± 0.09, which the team judged to be in good agreement with the result 0.72 ± 0.08 from the HST Key Project reported in 2001.

There could now be little doubt that dark energy accounts for about two-thirds of the total energy density of the universe, with both cold dark matter and visible baryonic matter accounting for one-third.

WMAP

Fifteen years passed between NASA's 'Announcement of Opportunity' in 1974 and the launch of the COBE satellite in 1989. Of course, the scientists were now on much firmer ground. It was clear that the temperature fluctuations were observable and the ground-based and balloon-borne experiments that had been conducted since the COBE mission had demonstrated that the acoustic peaks in the TT power spectrum were also observable. This greater certainty meant that timescales could now be compressed. MAP was approved for development in 1997, and launched from the Kennedy Space Center in Florida on 30 June 2001, aboard a Delta II rocket. The time between submission of competitive proposals and launch was just six years.

After a three-month journey, MAP was parked in a rather unique orbit, at a 'Lagrange point' in the Sun–Earth system, named for the Italian-French mathematician Joseph-Louis Lagrange. Typically, the gravitational pull exerted on an object such as a satellite by two large bodies (such as the Sun and the Earth) is unbalanced, dramatically altering the satellite's orbit. However, there are a number of so-called *Lagrange points* that can be identified at which gravitation and the centripetal force experienced by the satellite are much more balanced. The Lagrange point L2 lies on a line drawn from the Sun through the Earth, at a distance of 1.5 million km from the Earth (Fig. 37). A satellite positioned at this point orbits the Sun with the same orbital period as the Earth. With the Sun, Earth, and Moon all tucked away behind it, it has a clear uninterrupted view of deep space.

MAP actually described a small, six-month orbit around L2, but this was still an unstable position, and the satellite required course and altitude corrections every 23 days or so to maintain orbit. But the advantage of L2 is that the satellite is shielded from the Sun and from Earth's own magnetic field and microwave emissions, yet close enough to Earth to ensure easy communications. The on-board observatory consisted of two telescopes each comprised of cooled differential microwave radiometers capable of operating at one of five different frequencies and with 45 times the sensitivity of COBE, and primary reflectors designed to improve angular resolution by a factor of 33. These were rotated every two minutes.

MAP completed its first full-sky observations in April 2002 and a second by August, after one full year at L2. On 5 September, after a 17-year battle with cancer, David Wilkinson died in Princeton.

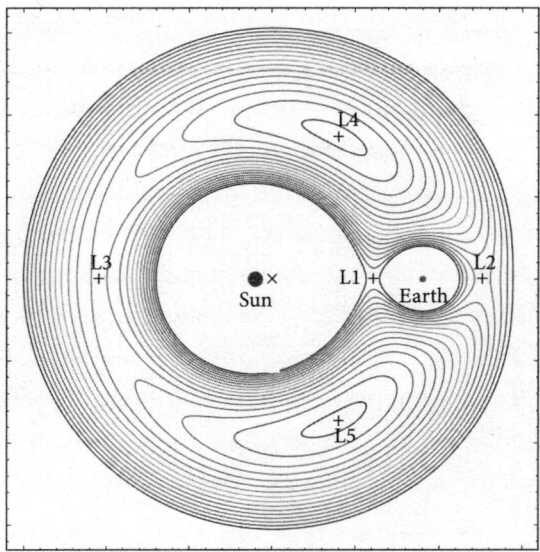

Fig. 37 This gravitational contour map of the Sun–Earth system shows the locations of five Lagrange points. The Lagrange point L2 lies 1.5 million km from Earth on the side facing away from the Sun.

Wilkinson had begun his student career as an engineer, earning bachelor's and master's degrees at the University of Michigan before turning to physics for his PhD. He had joined Dicke's research group at Princeton in 1963, just in time to get caught up in the scientific thrill-ride that was the discovery of the cosmic background radiation. His scientific career was thereafter dedicated to the study of this relic from the primordial universe: 'He set the learning curves on how to make those difficult measurements, and found the best places to do them: Princeton rooftops, deserts, mountains, balloons, and space', wrote John Mather, Lyman Page, and Jim Peebles in their obituary.[11] 'The results have driven the development of the standard model for cosmic structure formation and the new generation of cosmological tests. In the process of those experiments, Dave trained a large fraction of the scientists now engaged in this wonderfully productive field of experimental cosmology'.[12]

Photo 18 David Wilkinson died in Princeton on 5 September 2002, after a 17-year battle with cancer. In February 2003, the MAP satellite mission was renamed in his honour as the Wilkinson Microwave Anisotropy Probe.

On 11 February 2003, NASA Headquarters in Washington, DC issued a press release. The MAP satellite mission was to be renamed the Wilkinson Microwave Anisotropy Probe, in honour of Wilkinson's memory and scientific legacy. This was shortly followed by a second press release on the same day, jointly from NASA and the Goddard Space

Flight Center. The results from WMAP's first year of data collection were now available. A total of 13 papers were submitted that day to the *Astrophysical Journal*, subsequently published in a special supplement in September.

Public attention was inevitably drawn to the all-sky map of the background radiation, shown in Fig. 38(a). The improved sensitivity and angular resolution compared with COBE was strikingly apparent (compare this with Fig. 29). And, although the TT power spectrum didn't extend beyond what had been achieved by experiments such as BOOMERANG, the data were considerably more precise, Fig. 38(b).[13]

Such was the quality of the data that fitting the TT power spectrum was now possible using only six parameters, without the need for constraints from other studies. However, the precision of these parameters was improved by adopting the concordance approach, using data from other studies of the cosmic background at small angular scales, the results of a two-degree field galaxy redshift survey (2dFGRS) conducted by astronomers at the Australian Astronomical Observatory (to which we will return shortly), and studies of the 'Lyman-alpha forest' (Lyα forest), which provided information on the interstellar medium using light from high redshift quasars.

These 'fitted' parameters, and parameters derived from them of relevance to our story, are summarized in Table 2 below. Because the highest precision was derived by using priors from other studies, I've called these 'concordance results'. Of the fitted parameters, three are unfamiliar. The spectral index, n_s, is a measure of the scale invariance of the temperature fluctuations, with $n_s = 1$ signifying complete scale invariance. The parameter A is an amplitude factor related to the power spectrum, and the reionization optical depth, τ (Greek tau), is a measure of the fraction of cosmic background photons that is scattered from the line of sight of the microwave detector.

The WMAP data firmly ruled out the possibility of warm dark matter. The preferred cosmological model that emerged from this analysis was therefore termed the Λ-CDM (lambda, cold dark matter) model, now

(a)

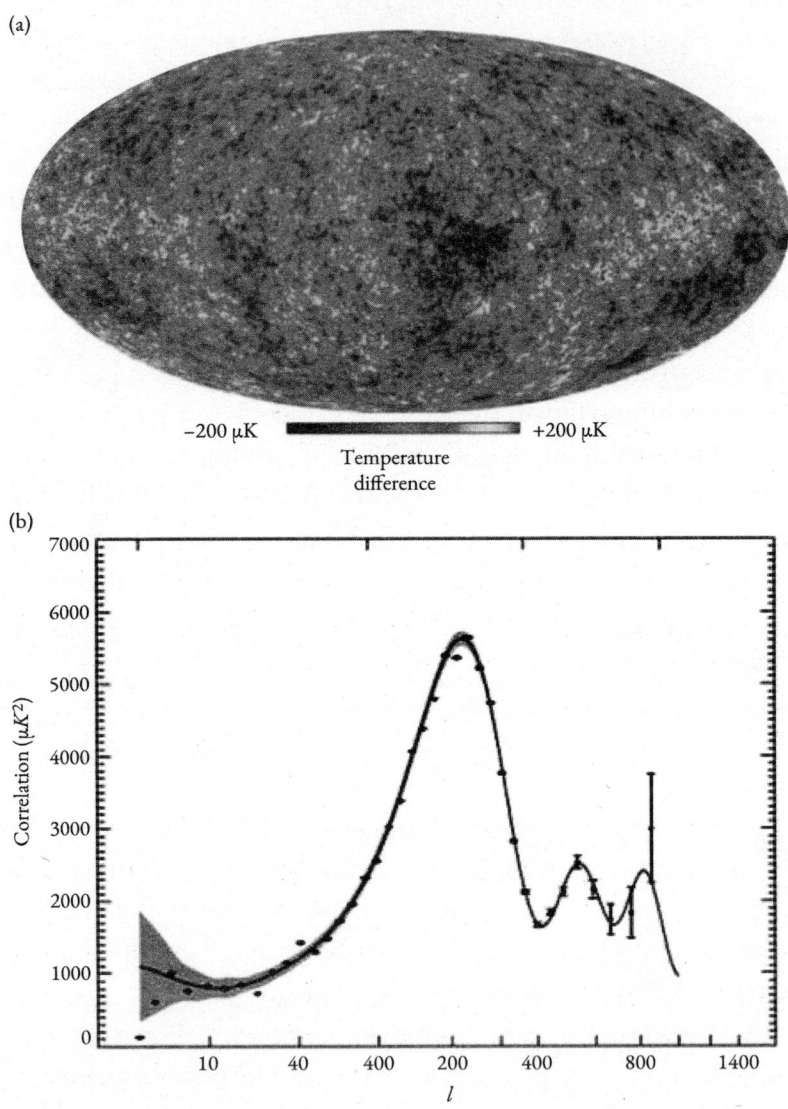

(b)

Fig. 38 The WMAP first-year results showed much greater sensitivity and angular resolution compared with COBE, reflected in both the all-sky map, (a), and the TT power spectrum, (b). In (a) the temperature differences are colour-coded and range from −200 millionths of a Kelvin (μK) (dark) to +200 μK (light). In (b) the continuous line through the data points is the best-fit Λ-CDM model.

Table 2. WMAP first-year concordance results (2003)

Fitted Parameters		Derived Parameters	
$\Omega_B h^2$	0.0224 ± 0.0009	Ω_B	0.044 ± 0.004
$\Omega_M h^2$	$0.135^{+0.008}_{-0.009}$	Ω_M	0.27 ± 0.04
h	$0.71^{+0.04}_{-0.03}$	$(\Omega_\Lambda$	$0.73 \pm 0.04)^a$
Spectral index, n_s	0.93 ± 0.03	Age, billion years	13.7 ± 0.2
Amplitude, A	$0.83^{+0.09}_{-0.08}$	Recombination, years	$379,000^{+8000}_{-7000}$
Optical depth, τ	0.17 ± 0.06	Sound horizon, Mpc	147 ± 2

[a] The WMAP team did not include Ω_Λ in their list of derived parameters. However, the team combined the WMAP data with results from other cosmological studies to determine $\Omega = 1.02 \pm 0.2$ (a flat universe), from which follows the estimate for Ω_Λ included in the table.

sometimes referred to as the 'standard model' of inflationary Big Bang cosmology. It is also sometimes referred to as the 'concordance model', but I would like to continue to distinguish here between the theoretical model itself (Λ-CDM) and the 'concordance approach' which uses results from multiple sources to fit the data derived from (for example) the TT power spectrum. The aim is to test the extent to which the Λ-CDM model can comfortably accommodate a set of values for the cosmological parameters based on many different ways of determining these: a 'concordance' which in one dictionary definition means 'agreement or consistency'.

The agreement between the WMAP concordance value for h of 0.71 ± 0.04 and the HST Key Project result for H_0 of 72 ± 8 km/s/Mpc must have been gratifying, and no doubt encouraged cosmologists to believe that further improvements in precision would lead ultimately to complete convergence. The significance of this could not be overstated. Distance ladder measurements provide a relatively direct determination of cosmological parameters such as H_0 from studies of 'nearby' objects with look-back times up to billions of years. Even when constrained by priors from other sources, analyses of the correlations in the temperature fluctuations of the cosmic background rely on the *assumption* of a cosmological model (such as Λ-CDM), and reach much further back in time. Agreement between the values of H_0 so derived would lend enormous support to the conclusion that the Λ-CDM model applies across the entire evolutionary history of the universe.

Writing in 2020, Peebles reflected: 'The change in the state of empirical cosmology in the five years from 1998 to 2003 was great enough to be termed a revolution'.[14]

10

The Hubble Tension

For many years astronomers and cosmologists had worried that the absence of observable temperature fluctuations in the cosmic background was inconsistent with the existence of stars and galaxies, and of large-scale structures of clusters of galaxies and voids in our present-day universe. Once the temperature fluctuations had been detected and mapped, and the acoustic oscillations had been teased from the TT power spectrum, it was inevitable that further questions would be asked of the baryonic matter that underpinned them. The temperature fluctuations represent anisotropies in the distribution of baryonic matter (the visible stuff of stars and galaxies) during the period of recombination. Theoretical calculations suggested that essential features of the large-scale structure of the universe should be preserved through its subsequent evolution. This implied that the acoustic oscillations—specifically the sound horizon—should be imprinted on the distribution of visible matter, frozen in place when matter and radiation parted company, as illustrated in Fig. 33.

Just as with the temperature fluctuations in the cosmic background, the imprint of the sound horizon on today's universe can only be observed by looking at *correlations* in the separation distances of galaxies. Given a random galaxy in a particular location in the sky, such a correlation is a measure of the probability that another galaxy will be found a specific distance from it. A slightly higher correlation between galaxies separated by about 150 Mpc is therefore evidence of the legacy of baryon acoustic oscillations imprinted on the large-scale structure of the universe. Observing these would provide an important test of Λ-CDM cosmology, as well as constraints on key cosmological parameters independently of studies of the cosmic background.

Baryon Acoustic Oscillations

Finding correlations in the separation distances of galaxies meant conducting further galaxy redshift surveys. The CfA redshift survey had begun in 1977, and CfA2 had reported redshifts of 15,000 galaxies by the early 1990s. Advances in technology now enabled much more efficient collection of spectroscopic data. The Las Campanas redshift survey (in which Kirshner was a team member) made use of a multi-fibre spectrograph mounted at the prime focus of the 2.5 m du Pont Telescope at Las Campanas Observatory in Chile's Atacama Desert. The Las Campanas survey recorded almost 24,000 redshifts up to $z = 0.2$, corresponding to recession speeds of 60,000 km/s, and look-back times of about 2.5 billion years (compare this with Fig. 27(a)).[1]

A UK initiative to convert existing telescopes to wide-field astronomy led fortuitously to an order of magnitude enhancement in survey efficiency. This was realized by the 2-degree Field Galaxy Redshift Survey (2dFGRS) using the 3.9 m Anglo-Australian Telescope operated by the Australian Astronomical Observatory on Siding Spring Mountain near Coonabarabran, New South Wales. The multi-fibre spectrograph used by the 2dFGRS team allowed the measurement of 400 spectra simultaneously over a 2-degree diameter field of view, about 16 times the area of the full Moon. Established in 1997, the 2dFGRS set out to determine the redshifts of up to about 250,000 galaxies.[2]

It soon had competition. The non-profit Astrophysical Research Consortium (ARC) was formed in 1984 to support the building of a new observatory at Apache Point in the Sacramento Mountains in Sunspot, New Mexico. This was funded by a consortium of universities whose astronomers would be allocated telescope time based on the size of each institutions' financial contribution. ARC's mission had many elements, but one of these was to break down the monopoly enjoyed by California's Palomar, Mount Wilson, and Lick Observatories. The ARC 3.5 m telescope was dedicated in May 1994, along with another 1 m telescope operated by New Mexico State University. At the time the ARC 3.5 m telescope was the 14th largest in the world.[3]

But just as these instruments were being dedicated, there was another telescope under construction at Apache Point. Ideas for a telescope dedicated solely to cosmology had bounced around among ARC members since the early 1980s. With support from the Alfred P. Sloan Foundation (established in 1934 by Alfred P. Sloan, Jr, then President and CEO of General Motors), another consortium was formed under ARC's umbrella. This is the Sloan Digital Sky Survey (SDSS) which uses a dedicated 2.5 m telescope to conduct, among other things, a comprehensive survey of galaxy redshifts with the aim of elucidating the large-scale structure of the universe.[4] Like Las Campanas and 2dFGRS, the SDSS exploits a multi-fibre spectrograph and a multi-array scanning CCD camera. The SDSS began collecting data in 2000.

Pickering had been obliged to confront the challenge posed by improved efficiency in the gathering of data. Now astronomers of the new century were confronted with the challenge of managing and analysing unprecedented large volumes of spectroscopic data. Pickering's solution had been the Harvard computers, and the solutions adopted by 2dFGRS and SDSS were modern (but non-human) equivalents based on large digital databases and computer software such as Structured Query Language (SQL). Teasing out evidence for baryon acoustic oscillations in the distribution of galaxies would require careful analysis of the data, and astronomers as much at home with computer software and databases as they were with telescopes.

Typical of the new breed of computational astronomers is Shaun Cole, at Durham University in England, a specialist in the processing and analysis of galaxy redshift data. He had joined the Anglo-Australian 2dFGRS team at its inception.[5] John Peacock at the University of

Edinburgh in Scotland (and principal clarinet with the Scottish Sinfonia) was a founder member of the Virgo Consortium for Cosmology Supercomputer Simulations in 1994. He became the UK chairman of the 2dFGRS team in 1999, joining Matthew Colless, his Australian co-leader, based at the Australian National University in Canberra.[6]

The SDSS team had agreed that determining large-scale structure one galaxy at a time was quite inefficient. What they needed to do was identify galaxy clusters, which meant writing software to pick them out from the stream of data gathered each night by the 2.5 m telescope at Apache Point. One characteristic of a galaxy cluster is unusually large, typically elliptical, bright red galaxies sitting at its centre. Find these and you could virtually guarantee finding clusters of galaxies surrounding them. At first the SDSS's Cluster Working Group called them big red galaxies, but Daniel Eisenstein, a post-doc at the Institute for Advanced Study in Princeton and specialist in statistical methods, preferred the name luminous red galaxies (LRGs).[7] Luminosity is intrinsic, but brightness depends on distance. A highly luminous galaxy may appear quite dim because it lies far away.

The 2dFGRS team experienced slow progress in its first two years of operation, gathering redshift data for only 50,000 galaxies. But the team had achieved 100,000 redshifts a year later. The survey was completed in April 2002, with a total of 221,283 galaxy redshifts. We get a sense for how far this kind of astronomy had come in Fig. 39, which compares the iconic results of the CfA2 survey with those of Las Campanas and 2dFGRS. SDSS produced similar results, and in 2003 the Sloan Great Wall was identified—lying about a billion light-years away (307 Mpc), measuring almost 1.5 billion light-years (460 Mpc) in length, more than twice as long as the Great Wall observed in the CfA2 survey.[8] The large-scale structure was clear: a complex web of 'large, almost empty voids surrounded by relatively thin sheets, intersecting in long filaments that in turn meet at dense nodes—galaxy clusters'. These structures are more visible at smaller scales 'reflecting the fact that the universe is nearly homogeneous on the largest scales'.[9]

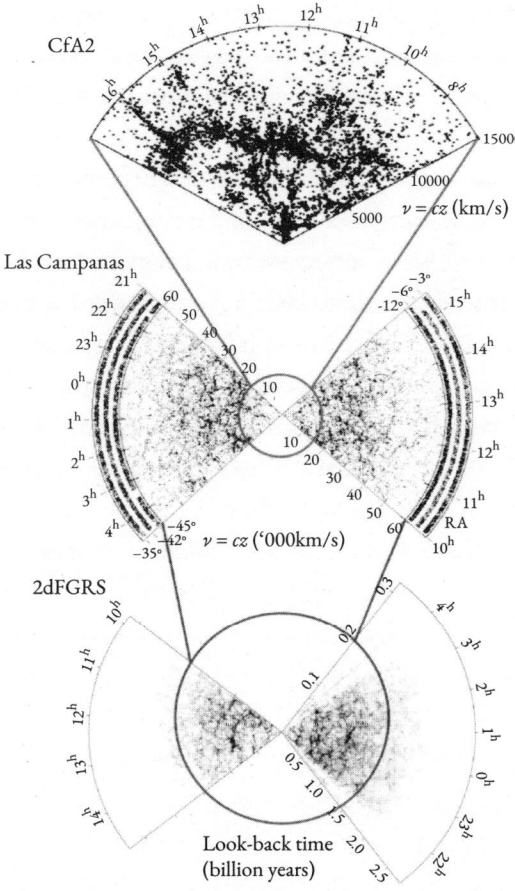

Fig. 39 Comparison of the results of the CfA2, Las Campanas, and 2dFGRS galaxy redshift surveys.

The 2dFGRS team published their analysis of the matter power spectrum—the baryonic matter equivalent of the TT power spectrum—in September 2005.[10] Fitting the redshift data directly to the Λ-CDM model assuming a spectral index $n_s = 1$ and a Hubble constant $h = 0.72$ yielded $\Omega_M h = 0.168 \pm 0.016$ and $\Omega_B / \Omega_M = 0.185 \pm 0.046$. These results, which were found to be relatively insensitive to the chosen value of h, return the values $\Omega_B = 0.043 \pm 0.051$ and $\Omega_M = 0.233 \pm 0.022$.

It is perhaps worth reminding ourselves that these estimates for the cosmological parameters are derived not from the cosmic background, but from studies of the large-scale distribution visible matter in the universe.

The SDSS team published their analysis of the two-point correlation in 'redshift space' a month later, based on a sample of 46,748 LRGs, up to $z = 0.47$.[11] This revealed a small but perfectly formed acoustic peak, equivalent to the last (and largest) peak in the TT power spectrum, at a separation distance of about $100/h$ Mpc (Fig. 40). A Hubble constant of $h = 0.7$ places this peak at 143 Mpc, corresponding to the sound horizon. This was a dramatic confirmation of Λ-CDM cosmology.

Eisenstein, Cole, and Peacock were awarded the 2014 Shaw Prize for astronomy, established in 2002 by Run Run Shaw, a film and television mogul based in Hong Kong. Previous winners include Peebles (2004),

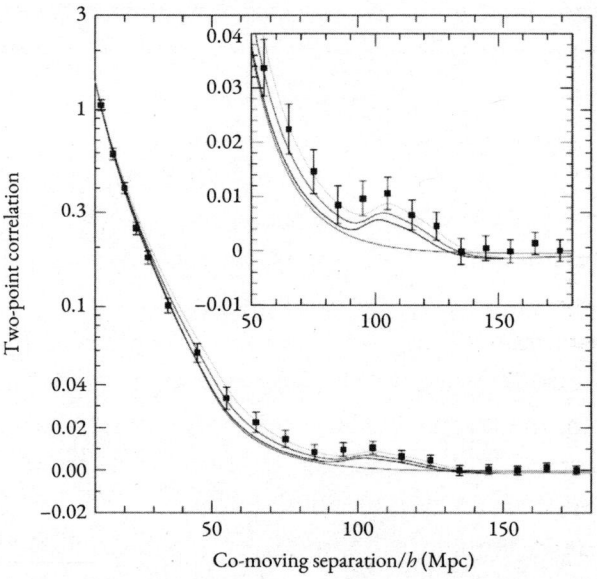

Fig. 40 The SDSS identified a peak in the two-point correlation of galaxy separation distances corresponding to the sound horizon. In this figure, the curves show predictions based on the Λ-CDM model with different estimates for $\Omega_B h^2$ and $\Omega_M h^2$. The inset shows an expanded view of the peak.

Perlmutter, Riess, and Schmidt (2006), and Bennett, Page, and Spergel (2010, for their work on WMAP).

A Standard Ruler

But the discovery of the acoustic peak also provided a potentially powerful new method to explore the properties of dark energy. The sound horizon is the result of the relatively simple physics of the radiation-dominated plasma occurring before recombination, and the acoustic oscillations were 'frozen' in place when radiation became disconnected from matter, about 380,000 years after the Big Bang. The position of the peak of the last compression wave is determined simply by the speed of sound in the plasma and the time elapsed between the Big Bang and recombination. It is a fixed characteristic of our universe, and the temperature fluctuations of the cosmic background provide a very precise determination of this position. It thus provides a 'standard ruler', unaffected by the subsequent expansion.

The two-point correlation shown in Fig. 40 puts the sound horizon at around $100/h$ Mpc, for galaxies with redshifts up to $z = 0.47$, corresponding to a look-back time of about 5 billion years. But suppose we now extend the galaxy redshift survey to $z = 3$, a look-back time of 11.5 billion years, before the expansion flipped from deceleration to acceleration. The position of the sound horizon will not have changed for these older galaxies, but the Hubble parameter will be different, as its value depends on the expansion history of the universe. Such extended redshift surveys could therefore be used to measure the Hubble parameter as a function of redshift, H_z (recall that the Hubble constant, H_0, is a measure of the Hubble parameter 'now', i.e., at $z = 0$). Aside from providing an independent measure of the accelerating universe, mapping the expansion history would also provide a sensitive probe of the dark energy that drives it.[12]

Attention turned to the dark energy 'equation of state', a term borrowed from thermodynamics. In a 'perfect' fluid, the equation of state is characterized by the ratio of the pressure of the fluid, p, to its mass density, ρ, according to $w = p/\rho$, where w is a dimensionless number. For the vacuum energy associated with Einstein's cosmological constant, the equation of state parameter is $w = -1$: the negative (anti-gravitational) pressure of the vacuum energy is related directly to its density. Determining an observational value for w would allow astronomers to discover if the dark energy driving the accelerating expansion of the universe is indeed equivalent to the vacuum energy associated with the cosmological constant. Any variation in the observed value of w with redshift or look-back time would be a powerful hint that there is more to dark energy than the cosmological constant (and the Λ-CDM model) suggests. A dynamic, rather than static, dark energy would be more indicative of the kind of quantum field that is thought to have driven cosmic inflation.

Eisenstein became the director of the third phase of the SDSS collaboration in 2008.[13] The largest of SDSS-III's four surveys, the Baryon Oscillation Spectroscopic Survey (BOSS), was launched in the same year.

SH$_0$ES

The multi-parameter fitting of the Λ-CDM model to the TT power spectrum and the redshift data from 2dFGRS and SDSS provided powerful constraints on the values of the cosmological parameters. But the analysis still required the assumption of a flat universe and dark energy as a cosmological constant with $w = -1$. Such analysis could be used to *predict* H_0 but Riess argued that it was still very necessary to make independent distance ladder *measurements* of H_0. He was inspired by an article by Wayne Hu at the University of Chicago, who wrote: 'The single most important complement to the [cosmic background radiation] ... is a determination of the Hubble constant to better than a few percent'.[14]

Riess judged that it was time to beat down the errors some more.

At issue was the perennial problem of accuracy and precision. Although the HST Key Project had been largely successful, the reported 11% error was now much larger than the error estimates emerging, for example, from the analysis of the WMAP data. The original WMAP project plan included two years of observations, but mission extensions were granted in 2002, 2004, 2006, and 2008, and the satellite had continued to gather more data on the cosmic background. The WMAP team had issued updated sets of results every two years. The WMAP 5-year observations (2009) reported a concordance result for H_0 of 70.5 ± 1.3 km/s/Mpc—an error of less than 2%—based on data from WMAP, baryon acoustic oscillations, and high-z Type Ia supernovae.[15] To make a meaningful contribution to the debate, it would be necessary to reduce the errors associated with the distance ladder measurements much further.

There are two broad sources of error in any observation or measurement. The first is random, or statistical. Repeated measurements tend to be subject to random fluctuations around a mean. If the bullseye is the 'right' answer, measurements subject to random error are scattered uniformly in and around the bullseye. The extent of the scatter determines the level of *precision* of the data. The second source of error is systematic, and is much more of a problem. Measurements subject to systematic error are skewed away from the bullseye, and affect the *accuracy* of the result. The measurements may be quite precise (not so much scatter), but as the mean no longer coincides with the bullseye, the result is inaccurate.

One of the largest sources of systematic error in previous distance ladder determinations of H_0 arose in combining different observations of Cepheid variables taken with different photometric systems. In July 2006, Riess bid for time on Cycle 15 of the HST to fix a few rungs and so 'refurbish' the distance ladder. He proposed to do this by studying Cepheid variables in galaxies that are also host to Type Ia supernovae, using the telescope's Near Infrared Camera and Multi-object Spectrometer (NICMOS). The aim was to improve the accuracy and precision of distance ladder measurements of H_0 by a factor of two, to less than 5%. These results were to be combined with further studies of supernovae with $z > 1$, allowing the possibility of distinguishing between static and dynamic dark energy. He called the project Supernovae and H_0 for the Equation of State of Dark Energy, or SH_0ES.

In her 1908 paper in the *Annals of Harvard College Observatory*, Leavitt had remarked that '... the brighter variables have the longer periods'.[16] A century later, the attentions of many astronomers returned to the Cepheid variables with a renewed sense of purpose. In June 2009, the SH_0ES team published results from studies of 240 long-period Cepheids in six galaxies hosting Type Ia supernovae.[17] The homogeneity of the Cepheids and the fact that they had been observed using the same instrument greatly reduced the systematic errors. The team obtained the result 74.2 ± 3.6 km/s/Mpc, an error of 4.8%. The 5-year WMAP determination of $\Omega_M h^2$ was used to deduce $w = -1.12 \pm 0.12$. Riess was confident that further improvements in precision were possible, reducing the error in H_0 to 1%.

The Planck Mission

The European Space Agency (ESA) was established in 1975 and in November and December 1983 issued invitations to the European scientific community to submit proposals as part of its Horizon 2000 programme. It received 30 proposals from astronomers. The COBE announcement in April 1992 prompted a surge of interest in the temperature anisotropy of the cosmic background. In response to a further invitation for proposals for a third round of medium-class space missions in May 1993, ESA received two competing but complementary proposals for new satellites. These were quickly combined to form the Cosmic Background Radiation Anisotropy Satellite/Satellite to Measure Background Anisotropies (COBRAS/SAMBA) mission. Fortunately, after clearing its preliminary assessments, it was renamed much more succinctly as the Planck mission in honour of German physicist Max Planck.

By February 1998, ESA had received proposals for two instruments to be placed aboard the Planck satellite. These were submitted by a team led by Jean-Loup Puget, director at the Institut d'Astrophysique Spatiale in Orsay, France, and a team led by Nazarreno Mandolesi, director of the Istituto TESRE/CNR in Bologna, Italy. These were rather unimaginatively called the Low Frequency Instrument (spanning frequencies from 30 to 70 GHz) and the High Frequency Instrument (100 to 857 GHz), with Mandolesi and Puget acting as Principal Investigators. The instruments were designed to provide 2.5 times the angular resolution available from WMAP, corresponding to $\theta \sim 0.07°$ and l values up to 2500.

Construction contracts were signed in June 2001, just as MAP was launched from the Kennedy Space Center. Jan Tauber, the Planck project scientist, was quick to stress that Planck would provide at least a factor of 10 improvement in sensitivity compared with MAP and should be considered as the third (after COBE and MAP), and most sophisticated generation of satellite mission. 'There is a progression, not competition'.[18]

The Planck satellite was scheduled for launch in February 2007, then August 2007, then February 2008, then July 2008. It was finally flown to the Guiana Space Centre, to the northwest of Kourou in French Guiana, on 19 February 2009, and prepared for launch aboard an Arianne 5 rocket. It would be accompanied by a second payload: the ESA's Herschel Space Observatory, then the world's largest infrared space telescope. After another short delay, Planck and Herschel were launched on 14 May 2009, and Planck was placed in orbit around the Lagrange point L2 on 3 July.

There had been no further mission extensions for WMAP beyond 2008, and in October 2010 the satellite was parked in a 'graveyard' orbit. The WMAP 7-year observations were published in 2011, with concordance results based on WMAP, baryon acoustic oscillations, and the value for H_0 that the SH_0ES team had reported in 2009.[19] The concordance result for H_0 held steady at $70.4^{+1.3}_{-1.4}$ km/s/Mpc.

The SH_0ES team now took advantage of the Wide Field Camera 3, installed on the HST during a space shuttle servicing mission on the same day that Planck had launched, to study 600 Cepheid variables in galaxies also host to eight recent Type Ia supernovae. The results were combined with calibrations based on 13 Cepheids observed in the Milky Way for which parallax measurements (and hence accurate distances) were available, and 92 Cepheids in the Large Magellanic Cloud for which distances derived from eclipsing binary systems were available. This combination allowed a best estimate for H_0 of 73.8 ± 2.4 km/s/Mpc, a 3.3% error.[20] Although it was too soon to be definitive, it appeared that within their respective ranges of errors, the concordance predictions for H_0 from studies of the cosmic background and large-scale structure were beginning to part company with the results of distance ladder measurements.

The divergence was compounded by the WMAP 9-year observations, published in 2013. The concordance results were based on WMAP, baryon acoustic oscillations, and the value for H_0 that the SH_0ES team had reported in 2011. Also included were temperature data from studies of the cosmic background by two ground-based telescopes—the Atacama Cosmology Telescope (ACT), high in the Chilean Andes above the Atacama Desert, and the South Pole Telescope (SPT), based at the Amundsen–Scott South Pole Station in Antarctica. Both telescopes had seen first light in 2007, and had since undergone substantial upgrades. In 2011 the ACT team had used its own 2008 survey data to produce a concordance estimate for H_0 of 69.9 ± 1.4 km/s/Mpc.[21] The WMAP team now returned the favour, producing a further estimate for H_0 of 69.32 ± 0.80 km/s/Mpc, an error of a little more than 1%.[22]

It is perhaps worth emphasizing once more that using parameters derived from other studies, including more direct distance ladder measurements of H_0, as priors in the analysis of the TT power spectrum does not mean that these inputs emerge unaffected by the fitting process. The priors serve as 'starting values' for the analysis, which finds its own set of six 'best fit' parameters from which H_0 is then calculated. Consequently, the value of H_0 obtained in this way remains an 'early-universe' *prediction* of the Λ-CDM model.

Planck had completed its first all-sky survey of the cosmic background in early 2010, shortly after ESA had approved a mission extension of 12 months. As had been expected, the coolant used to maintain the temperature of the High Frequency Instrument just 0.1 K above absolute zero ran out in January 2012. The instruments had completed five full surveys. Although the Planck collaboration had published preliminary results in 2010 and 2011, a first full set of results were submitted for publication in March 2013. The resolution of the temperature fluctuations in the all-sky map was quite extraordinary (Fig. 41(a)), and the TT power spectrum now revealed seven acoustic peaks (Fig. 41(b)).[23] 'Although I know that this is what modern cosmology is about', declared Puget, 'I am still amazed that this can really be done'.[24]

But the tension was now building. The Planck 2013 concordance results were based on Planck data for angular scales corresponding to $l = 50 - 2500$, combined with data from ACT ($l = 540 - 9440$) and SPT ($l = 2000 - 10,000$), baryon acoustic oscillations, and low-l data from WMAP that provided a slightly tighter constraint on the reionization optical depth, τ. The results are summarized in Table 3.[25] The unfamiliar parameter $100\theta_{MC}$ is 100 times the angular size (in radians) of the sound horizon at the time of recombination. The pattern was now set. For data derived from the early universe, the concordance analysis with the smallest error pointed to lower values for H_0. This trend continued with the Planck 2015 and 2018 results, which produced 67.74 ± 0.46 and 67.66 ± 0.42 km/s/Mpc, respectively.[26]

The Planck 2018 results set the standard for our present understanding of the cosmological parameters and Λ-CDM cosmology. Visible baryonic matter—the stuff of planets, stars, and galaxies—accounts for just 4.9% of the total mass–energy of the universe. The mysterious dark matter, an outrageous speculation from 1933 that 40 years later was

(a)

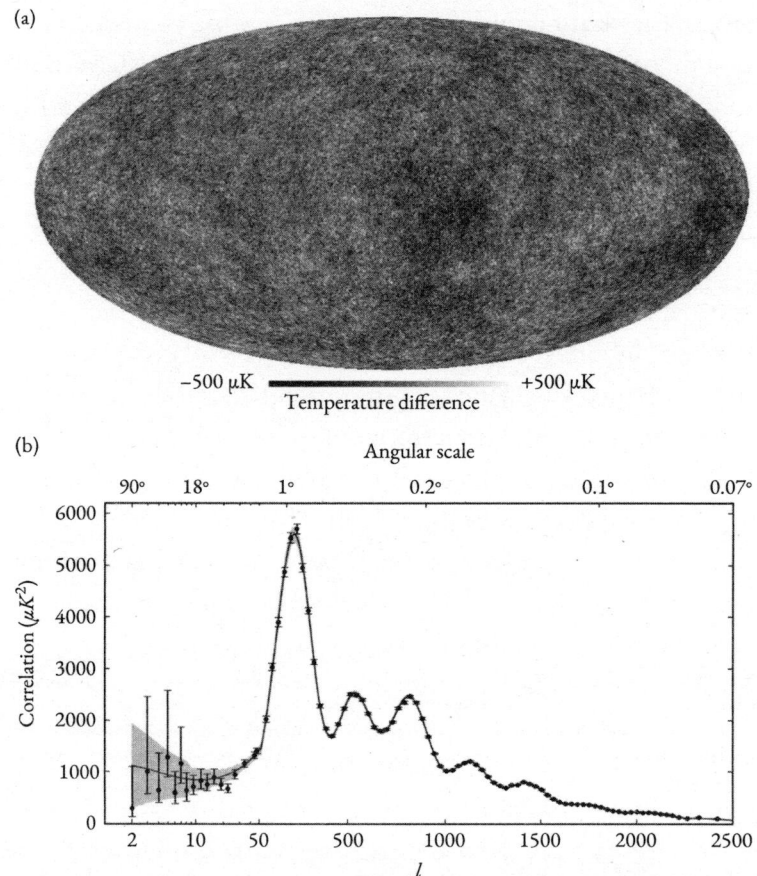

−500 μK ▬▬▬▬▬▬▬ +500 μK
Temperature difference

(b)

Fig. 41 The Planck collaboration released its first set of results in 2013. The all-sky map, (a), now showed much finer temperature variations than its predecessor, WMAP, and the TT power spectrum, (b), now revealed seven acoustic peaks.

found to be necessary to understand how galaxies form and to explain the pattern of rotational speeds of the stars within them, accounts for 26.1%. So, between them baryonic and dark matter account for 31.0% of the universe, just under a third. Finally, the even more mysterious dark energy, powering the accelerating expansion, accounts for 69.0%, consistent with a flat universe. With a dark energy equation of

Table 3. Planck 2013 concordance results

Fitted Parameters		Derived Parameters	
$\Omega_B h^2$	0.02214 ± 0.00024	Ω_Λ	0.692 ± 0.010
$\Omega_{DM} h^2$	0.1187 ± 0.0017	H_0, km/s/Mpc	67.80 ± 0.77
$100\theta_{MC}$	1.04147 ± 0.00056	Age, billion years	13.798 ± 0.037
Spectral index, n_s	0.9608 ± 0.0054	Sound horizon, Mpc	147.68 ± 0.45
$\ln\left(10^{10} A_s\right)$	3.091 ± 0.025		
Optical depth, τ	0.092 ± 0.013		

state parameter determined to be $w = -1.03 \pm 0.03$, it seems that dark energy is indeed the vacuum energy associated with Einstein's cosmological constant, and which had been a key component of Lemaître's fireworks universe.

In 2018, Mandolesi, Puget, and the Planck team were awarded the Gruber Prize for Cosmology, established in 2000 by The Gruber Foundation, organized by Yale University and funded by The Peter and Patricia Gruber Foundation. Other Gruber prize recipients that have featured in this story include Sandage and Peebles (2000), Rees (2001), Rubin (2002), Guth and Linde (2004), Mather and the COBE team (2006), Schmidt and the High-Z team and Perlmutter and the SCP (2007), Mould, Kennicutt, and Freedman (2009), Bennett and the WMAP team (2012), Starobinsky (2013), Tully (2014), Ostriker (2015), and Silk (2019). Puget was also awarded the 2018 Shaw Prize.

Peebles, who had been right at the centre of so many extraordinary theoretical developments in cosmology spanning 45 years, was rewarded with a share in the Nobel physics prize for 2019. He took the opportunity afforded by his Nobel lecture to offer some criticism of the Nobel committee, which had failed to recognize Dicke's 'deep influence in the development of gravity physics and cosmology'. 'But I am satisfied now', he concluded, 'because my Nobel Prize is closure of what Bob set in

Photo 19 Jim Peebles had been right at the centre of many extraordinary theoretical developments in cosmology spanning 45 years. He was rewarded with a share in the Nobel physics prize for 2019.

motion, his great goal of establishing an empirically based gravity physics, by the establishment of the empirically based relativistic cosmology'.[27]

The Hubble Tension

Something was not quite right. Further beating down the errors in the determination of H_0 should have led to a convergence between 'early-universe' predictions based on analyses of the cosmic background and baryon acoustic oscillations and 'late-universe' distance ladder measurements. Here 'late' simply refers to the fact that such direct measurements are inevitably made on those objects that are more local to Earth, with look-back times that are much shorter and so much later in the evolutionary history of the universe. But from about 2009 onwards, as the precision increased the results began to diverge.

Whilst the first year and three-year WMAP determinations of H_0 were entirely consistent with the HST Key Project result of $\sim$ 72 km/s/Mpc, subsequent early-universe predictions have tended to favour values more like $\sim 67 - 68$ km/s/Mpc. These predictions include later WMAP results and the Planck results, together with results reported in 2017 by the SDSS-III's BOSS programme (67.3 ± 1.0),[28] and results reported in 2011, 2020 (67.6 ± 1.1), and 2023 $(68.1 \pm 1.0$ km/s/Mpc) by the Atacama Cosmology Telescope.[29]

In the meantime, the SH_0ES team was busy refining its local distance ladder measurements based on Cepheids. The Wide Field Camera 3 (WFC3) on board the HST provided for an improvement based on near-infrared observations of Cepheids, more than doubling the number of variables observed in galaxies also host to Type Ia supernovae. In 2016 the team published the result 73.24 ± 1.74 km/s/Mpc.[30]

WFC3 also offers an angular resolution of 20 to 40 millionths of an arcsecond, using a technique called optical spatial scanning. In 2014, Riess and his colleagues heralded the possibility of measuring stellar parallaxes—and hence distances—up to an unprecedented 5000 parsecs. This put many long-period Cepheid variables within the Milky Way within reach of parallax measurements, allowing greater precision for the bottom rung of the distance ladder through direct calibration of the period–luminosity relation. Including local long-period variables in this calibration allowed the team to extend the increased precision to extra-galactic Cepheids in galaxies also host to Type Ia supernovae. This is the second rung of the distance ladder, and any increases in precision in the first necessarily extend to the second.

The team demonstrated these possibilities by measuring the parallax, and hence distance, of the Cepheid SY Aurigae. Four years later, the team reported parallax measurements for a further seven Milky Way Cepheids, and augmented the SH_0ES 2016 data to produce, in March 2018, a revised value for H_0 of 73.48 ± 1.66 km/s/Mpc, a precision of 2.3%.[31]

ESA's Global Astrometric Interferometer for Astrophysics (Gaia) Space Observatory was launched in December 2013. One of its mission parameters is to measure the parallaxes of 1 billion stars (including Cepheids) within the Milky Way with an accuracy of 20 millionths of an arcsecond, rising to 200 millionths of an arcsecond for the brightest stars. Problems arose when it was discovered that the telescope flexes very slightly as it rotates with respect to the Sun, producing a wobble that looks just like parallax. But, as the scientists gathered data, it became possible to distinguish the 'fake parallax' from real, and devise a methodology that researchers could use to make their own corrections.

The Gaia Data Processing and Analysis Consortium has issued three data releases, in September 2016, April 2018, and an 'early' data release in December 2020 (EDR3). This last release includes improved positions, parallaxes, and proper motions of stars, and was eagerly anticipated. In a paper published in February 2021, Riess and his colleagues used parallax measurements of 75 Milky Way Cepheids from Gaia EDR3 combined with brightness measurements from the HST to refine further the bottom rung of the distance ladder. They derived a value for H_0 of 73.2 ± 1.3 km/s/Mpc, a precision of 1.8%.[32]

By July 2022, the SH_0ES team had extended the data set to include Cepheids in the host galaxies of 42 Type Ia supernovae, calibrated using Gaia EDR3 parallaxes. The logic runs like this. Parallax measurements are used to calibrate the distances and luminosities of Milky Way Cepheids. The luminosities of more distant Type Ia supernovae are then calibrated using Cepheid luminosities. The Hubble constant is then inferred from the luminosities of high-redshift supernovae. The team reported the result 73.30 ± 1.04 km/s/Mpc, a precision of 1.4%.[33]

These results are presented in Fig. 42. Note that my use of time (since January 2001) as an axis in this figure has no significance other than as a means of observing the increasing precision (smaller error bars) in the determination of H_0 as both measurement and analysis techniques have improved over the last 23 years.

The divergence appears quite stark, and significant. Although we might quibble that the difference between the Planck 2018 result of 67.66 and the most recent SH_0ES result of 73.30 km/s/Mpc is not that great in the larger scheme of things (or in the context of the Hubble constant's troubled history) we should pay attention to the error

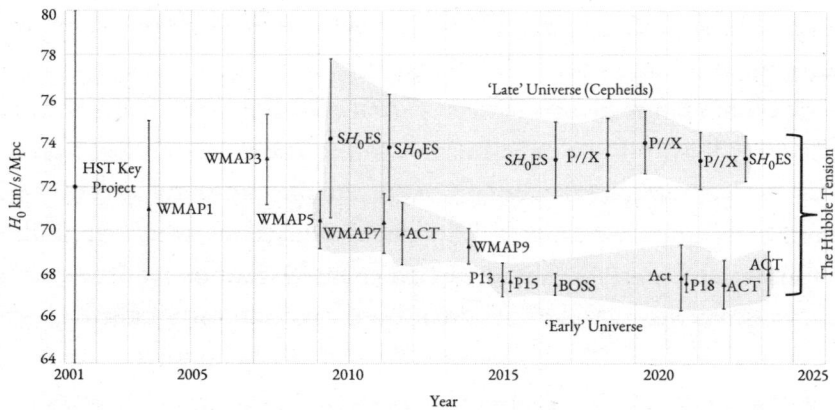

Fig. 42 Reported values of H_0 are here plotted against the time of publication spanning 22 years. As the precision of the reported values has increased (the 1σ error bars have reduced in size), the results from early-universe predictions and late-universe distance ladder measurements based on Cepheid variables and Type Ia supernovae have diverged, creating 'the Hubble tension'. The data labels P13, P15, and P18 refer to the Planck 2013, 2015, and 2018 results, respectively. P//X refers to parallax measurements using the HST WFC3 and data from Gaia EDR3. All other acronyms are explained in the text.

bars. These are the '±' ranges quoted with the results, and represent what statisticians call a single standard deviation, or 1σ (1-sigma) significance, providing 68% 'confidence limits'. If a confidence limit of little more than two-thirds doesn't sound very compelling, you might be reassured to know that in the social sciences a confidence limit of 95% (2σ) is usually required for a result to be accepted. And in high-energy particle physics, a confidence limit of 99.99994% (5σ) is required before the discovery of a new particle can be publicly declared. This is a big deal. Results with 5σ significance were required from both of the independent detector collaborations at the Large Hadron Collider before declaring that the Higgs boson had been discovered in 2012.

We can easily do the sums. A crude estimate of 5σ for the 2022 SH_0ES result is ±5.2 km/s/Mpc, implying a 'late' universe estimate for H_0 spanning the range 68.1 − 78.5 km/s/Mpc. The Planck 2018 result still lies just outside this range. *The universe is expanding faster than predicted by analyses of acoustic oscillations from the early universe.* This is the 'Hubble tension'.

Riess has compared the situation to a civil engineering project that has gone disastrously wrong. Imagine the construction of a (metaphorical) bridge spanning the age of the universe, begun simultaneously on both 'early' and 'late' sides of the divide. Foundations, piers, and bridge supports have been completed, but the engineers have now discovered to their dismay that the two sides do not quite meet in the middle.[34]

Revenge of the Acronyms

I have deliberately chosen the data points in Fig. 42 to highlight the Hubble tension at its most extreme, between early-universe predictions derived from concordance analyses using Λ-CDM cosmology and late-universe measurements based specifically on Cepheid variables and Type Ia supernovae. Although the SH_0ES team had employed various approaches to calibrating Cepheid distances, if nothing else history had taught astronomers not to rely exclusively on a single distance indicator.

Other distance ladder approaches have been tried but, with one exception, these have not yet approached the precision of the SH_0ES measurements. Mira variables (named, like Cepheids, for the prototype star Mira in the constellation Cetus) are red giant stars in the later stages of their evolution. Although they have less than twice the mass of the Sun, they are large enough to have passed through the helium burning stage in their cores and are thousands of times more luminous. There are broadly two subtypes, those rich in oxygen and those rich in carbon. The oxygen-rich Mira variables have a simple, linear period–luminosity relation, like the Cepheids but with longer periods typically spanning hundreds of days (Mira's period is 332 days).

A systematic search for Mira variables in the galaxy NGC 1559 (host to Type Ia supernova SN 2005df) using the HST's WFC3 turned up

115 oxygen-rich Miras. Calibrations borrowed from the SH_0ES programme (Riess was a co-author) yielded an estimate for H_0 of 73.3 ± 4.0 km/s/Mpc, published in January 2020.[35] Although Mira variables are harder to find and correctly identify than Cepheids, it is reasonable to assume that further data will serve to beat down the errors here too.

An approach based on observations of the fluctuations in the surface brightness of 63 galaxies up to distances of 100 Mpc yielded 73.3 ± 2.4 km/s/Mpc in April 2021.[36] In March 2020 the Harvard–Smithsonian CfA Megamaser Project made use of naturally generated water masers* in the accretion discs surrounding supermassive black holes to provide distance estimates to six galaxies. Among these was NGC 4258, an intermediate spiral galaxy in the constellation Canes Venatici whose distance of 7.576 ± 0.112 Mpc provides an important reference or 'anchor' point.[37] The Megamaser Project obtained 73.9 ± 3.0 km/s/Mpc.[38]

The first detection of gravitational waves in 2016 provided yet another approach. Gravitational waves are distortions in spacetime propagating away from the universe's most violent events, such as the merger of black holes and neutron stars. Analysis of the gravitational wave event GW170817 produced an estimate of 74^{+16}_{-8} km/s/Mpc.[39] The significant error arises, in part, from uncertainties related to the proximity of the event, which occurred at a distance of only 40 Mpc, too close to be free of the effects of peculiar motions. It is anticipated that an accumulation of data from events over the next five years could produce a determination of the Hubble constant with a precision of 2%.

In a discipline littered with acronyms, the award for the most contrived must surely go to H_0 LiCOW, standing for H_0 Lenses in COSMO-GRAIL's Wellspring.† The extent of the bending of starlight as it passes by the Sun during an eclipse, famously measured by Eddington and his

---

* Maser stands for microwave amplification by the stimulated emission of radiation. It is the microwave equivalent of a laser (where 'l' stands for 'light').

† COSMOGRAIL is yet another acronym and stands for Cosmological Monitoring of Gravitational Lenses.

colleagues, helped to confirm Einstein's general theory of relativity in 1919. Any massive object sitting in our line of sight—such as a galaxy—will likewise distort the light from more distant objects behind them. This effect is called gravitational lensing. Light from distant quasars can be distorted as it passes by an intervening galaxy, producing between two to four different images of the same object.

The COSMOGRAIL project, led by the École Polytechnique Fédérale de Lausanne, Switzerland, measures the time delays between individual images of the same quasar. These depend on the distances both to the lensing galaxy and the quasar. Modelling the distribution of mass in the lensing galaxy then provides access to the distance of the galaxy and thence H_0. In November 2016 the H_0LiCOW collaboration reported the result $71.9^{+2.4}_{-3.0}$ km/s/Mpc, and in September 2019 refined this to 73.3 ± 1.8 km/s/Mpc.[40] Another collaboration, called STRIDES (Strong Lensing Insights into Dark Energy Survey) used similar techniques to report $74.2^{+2.7}_{-3.0}$ km/s/Mpc in March 2020.[41]

The Tip of the Red Giant Branch

The implications are reasonably clear. Although these alternative approaches do not yet match the precision achieved by the SH_0ES collaboration based on Cepheids and Type Ia supernovae, they all point towards the higher values of H_0 that have become a characteristic of late-universe distance ladder measurements. They reinforce the idea of the Hubble tension.

Now we come to the exception.

The Tip of the Red Giant Branch (TRGB) approach makes use of the peak brightness reached by red giant stars after they stop fusing hydrogen and begin fusing helium in their cores. Recall that the individual stars within a single galaxy are observed at different stages of their evolution. A plot of luminosity against colour or spectral type (proxies for surface temperature) is an H–R diagram. Most of the stars within the population fall on the main sequence. The higher the surface temperature, the brighter the star shines and the bluer its colour. A star joins the main sequence at a position that depends on its initial mass, size, age, and evolutionary history. As it fuses the hydrogen nuclei (protons) in its core to form helium nuclei, it brightens and ascends the diagonal. A star the size of the Sun will spend about 10 billion years on the main sequence.

The pressure of the radiation produced by the nuclear fusion reactions helps to hold the star up against gravitational collapse. But when all the hydrogen nuclei are gone, the reactions and the radiation cease. Hydrogen nuclei in layers outside the core start to burn. The star swells to become a red giant, passes swiftly through the Hertzsprung gap, and follows the red giant branch typically to the right of the H–R diagram off the main sequence. The core contracts, driving up the temperature and pressure. For a relatively low mass star, the rising temperature triggers helium fusion reactions, releasing radiation once more. The temperature in the core starts to run away, increasing the rate of helium fusion and causing a dramatic increase in the luminosity of the star, called the *helium flash*. This happens at the tip of the red giant branch.

Now switch the logic of this description to the entire population of stars in a nearby galaxy which, at any moment of observation, will have a number of stars sitting at the tip. From among these choose older, low metallicity Population II stars with much less variability in luminosity and look at stars in the halo rather than the central bulge or disc of the galaxy as measurement of their luminosities is likely to be less affected by crowding among neighbouring stars. The H–R diagrams of such stars feature a prominent discontinuity—basically where the red giant branch ends. Observation reveals their apparent magnitudes (m) at the tip. Because the helium flash depends on the temperature and properties of the helium core, the absolute magnitude (M) of TRGB stars is predictable (it is ~ -4) and can be calibrated for instruments aboard the HST. The TRGB stars provide potentially powerful standard candles.

As with the Cepheids, this methodology is based on the observation of individual TRGB stars, and so is restricted to nearby galaxies. It nevertheless provides an important alternative to the Cepheids, whose physics are somewhat less certain and so less predictable. Distance determinations using TRGB stars can also be combined with observations of Type Ia supernovae to provide distance ladder measurements for H_0.

In 2015, In Sung Jang and Myung Gyoon Lee at Seoul National University in South Korea reported the results of their analysis of HST archival observations of TRGB stars in the outer regions of the Antennae galaxies NGC 4038 and 4039, a pair of interacting galaxies in the constellation Corvus, and NGC 5584, a barred spiral galaxy in the constellation Virgo. Measurements of the apparent magnitudes of the TRGB stars combined with absolute magnitude calibrations yielded distance moduli and hence distances for these galaxies.

Both were host to Type Ia supernovae in 2007, allowing the calibration of a magnitude–distance relation for the supernovae which was then extended to supernovae in three other nearby galaxies. The result was an estimate for H_0 of 69.8 ± 3.9 km/s/Mpc.[42] Two years later, Jang and Lee determined TRGB distances to a further three galaxies that were host to Type Ia supernovae in 1994, 1995, and 2002. Their revised best estimate for H_0 was 71.17 ± 1.87 km/s/Mpc.[43]

With the publication of the final HST Key Project results, Freedman figured that her part in the story of the Hubble constant was over: '... in 2001 I was convinced I'd never work on the Hubble constant again'.[44] But with colleagues from the Carnegie Observatories in Pasadena and a few other institutions in the US and Canada, in 2011 Freedman established the Carnegie Hubble Program. The aim of the programme was the determination of H_0 to a precision of 2%. An unlooked-for job offer proved too tempting to resist, and Freedman moved to the University of Chicago in September 2014, after 30 years at the Carnegie Observatories and 10 years as its director. Two years later the use of RR Lyrae variables and TRGB stars became the centrepiece of the Carnegie–Chicago Hubble Program (CCHP).

Photo 20 In 2001, Wendy Freedman had figured that her part in the story of the Hubble constant was over, but 10 years later she set up a new programme to determine H_0 with a precision of 2%.

The results followed a few years later. In September 2019, the CCHP team (including Jang and Lee in Seoul) reported $H_0 = 69.8 \pm 1.7$; in March 2020, 69.6 ± 1.7; and in September 2021, 69.8 ± 1.6 km/s/Mpc.[45] These results call into question the extent of the Hubble tension, which had by now provoked considerable speculation about the need to introduce 'new physics' beyond the Λ-CDM model to explain it. Freedman declared: 'The TRGB results alone do not demand additional new physics beyond the standard (Λ-CDM) cosmological model'.[46]

In November 2019 Riess and his colleagues challenged aspects of the calibration of TRGB stars and used the data to return a revised determination of H_0 more in keeping with the SH_0ES result.[47] The CCHP team responded with their March 2020 update. 'We run all of these tests, [and] keep on getting the same answer', explained Madore.[48] In March 2022, Freedman and Madore and their colleagues in their turn challenged the use of Gaia EDR3 parallaxes to calibrate Cepheid distances.[49]

The situation is confused. In 2022, reanalysis of TRGB data stored in the Extragalactic Distance Database (EDD),[50] managed by the University of Hawaii, was supplemented by new HST observations to yield an estimate for H_0 of 71.5 ± 1.8 km/s/Mpc.[51] Riess was a co-author. A 2022 pilot study involving TRGB calibration of the luminosity of Type Ia supernovae sought to reduce the systematic errors arising both from the host galaxies and photometrics. Freedman and Madore were co-authors. The study showed promise but its estimate for H_0 of 76.94 ± 6.4 km/s/Mpc was too imprecise to be informative.[52] In 2023, a team of astronomers including Riess made use of an algorithm to perform a Comparative Analysis of TRGBs (CATS), producing a baseline result of 73.22 ± 2.06 km/s/Mpc.[53] The team traced the differences with the earlier CCHP result to differences in the analysis of the Type Ia supernovae (accounting for 2.0 km/s/Mpc), and differences in the calibration of the TRGB stars (accounting for a further 1.4 km/s/Mpc).

Fig. 43 summarizes the state of play. Riess is convinced that the Hubble tension is real. Freedman is more circumspect, arguing in September 2021 that further studies of the Cepheids and TRGB stars, and the (then) anticipated launch of the James Webb Space Telescope (JWST) promise improvements sufficient to resolve the current discrepancies within a few years.

The JWST is a joint NASA, ESA, and Canadian Space Agency project. It was launched on Christmas Day, 2021, aboard an Arianne 5 rocket from Kourou in French Guiana, arriving at the Lagrange point L2 in January 2022. It is the largest optical telescope in space, designed to image near-infrared light, and offers three times the resolution and 10 times the sensitivity of the HST. The first images from JWST were released in July 2022. These are false-colour images produced from its near-infrared camera.

The JWST is already giving us more to say, and much more to think about.

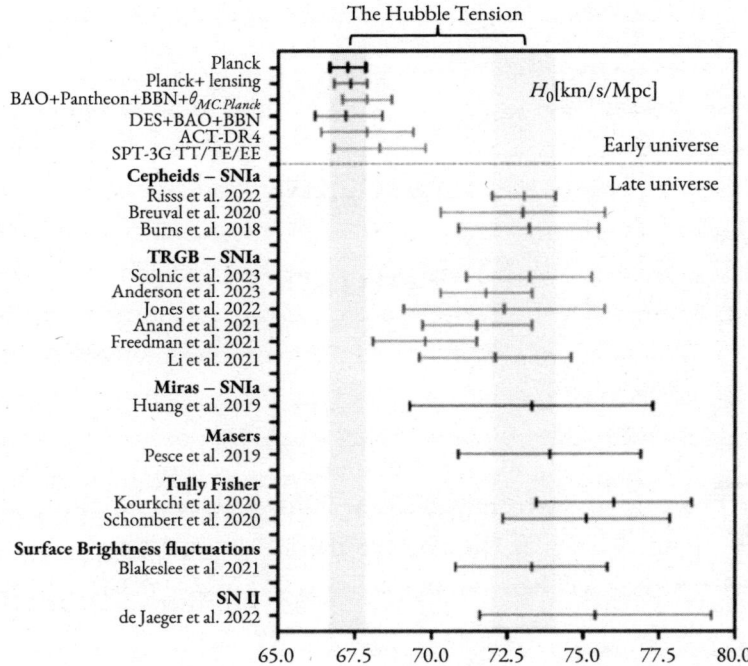

Fig. 43 Comparison of recent determinations of the Hubble constant derived from early-universe measurements of the cosmic background and from late-universe measurements based on a variety of distance indicators.

Epilogue

Discordance

Where do we go from here? One immediate task for astronomers is to discover if the Hubble tension is real. This means first resolving what I will here call the 'secondary' Hubble tension, between distance ladder measurements whose lowest rungs are based on different standard candles.

To be sure, the secondary Hubble tension does not signify a return to the Hubble wars of the 1970s. Then the arguments were about a factor of two difference in different estimates for H_0, against the background of an understanding of the large-scale structure of the universe that was a lot less sophisticated than it is today. The tension between the SH_0ES measurements based on Cepheid variables and the CCHP measurements based on TRGB stars is just 5%, and the results actually overlap at the level of about 2σ (95%) significance.

Astronomy long ago ceased to be just about *observing*. The raw observational data requires careful manipulation and analysis. Attention is inevitably turning to the usual suspects: calibration of the respective zero points, photometric errors, the reddening effects of dust, the effects of the age (and metallicity) of the stars used as standard candles, and the potential for bias resulting from crowding and blending of the astronomical images.[1]

Different standard candles offer different advantages and disadvantages. By their nature, Cepheids are younger stars that tend to be found on the galactic disc near to where they formed, where they are more likely to be surrounded by interstellar gas and dust, and more prone to crowding and blending effects. TRGB stars can be found both in the disc and the galactic halo, and careful selection means that their analysis can in principle be carried out with many fewer complications.

The relentless pursuit of more and more precise data will continue. At some time hopefully in the next few years we should see the secondary Hubble tension disappear. Then we will know if the primary tension between early-universe predictions and late-universe measurements is something we should be really worried about.

The Promise of the JWST

The JWST is already at the centre of new developments. One of its key mission goals is to fill in the gap between our early-universe observations of the cosmic background and our later-universe observations of high-redshift galaxies and nearby structures. Recall that the cosmological redshift associated with the cosmic background is $z \sim 1100$, a look-back time of 13.1 billion years, and the HST's serendipitous observation of SN 1997ff set a new galaxy redshift record of $z = 1.7$, with a look-back time of 9.7 billion years.

Quite a lot is supposed to have happened in the ~ 3 billion years in between. Following recombination, 380,000 years after the Big Bang, the light that was released became further redshifted as the universe continued to expand. The universe went dark. The matter in it consisted largely of dark matter and hydrogen and helium atoms, and a small scattering of slightly heavier elements. It is thought that dark matter slowly gathered itself together to form halos which in turn helped to gather and concentrate the baryonic matter at their centres. This allowed the very first—inevitably called Population III—stars to form, lighting up the universe once more, bringing the 'Dark Ages' to an end and assembling the first galaxies. These stars are hypothetical, and none has been confirmed so far. But there are strong logical and theoretical grounds for thinking that they provided an important stepping stone to the Population II and Population I stars we know today.[2]

If they did form, then these first stars would likely have been very massive, the temperatures and pressures at their cores causing them to burn through their hydrogen fuel very quickly. They would have been much larger and much more luminous than the stars we see today, glowing with predominantly ultraviolet light. They would have lived briefly but spectacularly, with an average lifetime thought to be just a few million years, their powerful light stripping electrons from the neutral hydrogen atoms in interstellar clouds to form free protons and electrons once more, in a process called reionization.

It is not known when these first stars and galaxies formed, or when reionization happened, but the odds favour a period corresponding to redshifts between $z = 20$ and $z = 6$, or 200 to 900 million years after the Big Bang. As the galaxy redshift record continued to tumble—$z = 4.9$ (1998); $z = 6.6$ (2003); $z = 8.2$ (2009); $z = 11.1$ (2016)—the extent of redshifting placed greater and greater emphasis on observations in the near infrared. This is why the instruments on board the JWST include a near infrared camera (NIRCam) and a near infrared spectrograph (NIRSpec).

In the short time that the instruments aboard the JWST have been active, the previous redshift records have been blown out of the water. The JWST Advanced Deep Extragalactic Survey (JADES) has focused on an area in and around the HST's ultra deep field, containing almost 100,000 galaxies in an image the size of a human seen from a mile away. Ten days of mission time with NIRCam, followed by three days of observing with NIRSpec, have allowed the study of 250 faint galaxies. Among them is JADES-GS-z13-0, which has a redshift $z = 13.2$, its light reaching back to about 300 million years after the Big Bang.[3]

In July 2024, an extremely red galaxy, JADES-GS-z14-0 (named Cerberus), was reported in the HST's ultra deep field. Limited data prevented a definitive identification, but one possibility seriously considered by the authors is that Cerberus is a large galaxy with a redshift $z = 15$, a look-back time of 13.2 billion years, and that its light was emitted just 260 million years after the Big Bang.[4]

A parallel Cosmic Evolution Early Release Science Survey (CEERS) aims to provide the community with data from the instruments on board the JWST pertaining to the formation of the first stars and reionization, and the assembly of galaxies.[5]

These relatively large, fully formed galaxies and galaxy clusters are appearing at some disconcertingly early times, earlier than anticipated from computer simulations which rely on the shepherding effects of dark matter, the formation of Population III stars (thought to take about 200 million years), the hierarchical assembly of these stars into large galaxies through the merging of smaller galaxies (taking at least a further 100 million years or more), and their accumulation in galaxy clusters. The results of computer simulations depend to a significant extent on how the calculations are set up, and are no substitute for empirical data. It is therefore too soon to tell if the observations from the JWST are truly at odds with Λ-CDM cosmology, but there are grounds for considerable uneasiness. We may yet be experiencing a new age paradox.

Although measuring the Hubble constant is not a JWST key mission goal, the data it produces is already informing the debate. The JWST's near infrared capabilities offer the promise of sharper observations free of the distorting effects of dust. In their relentless pursuit of more data, both Riess and Freedman have been allocated mission time to repeat their distance ladder measurements. 'As part of its legacy, HST resolved the factor-of-two debate in the Hubble constant, but even with two additional decades of progress, outstanding uncertainties still remain. A legacy of JWST will be the resolution of the current tension, and a robust answer to this question: "Is there new physics required beyond the standard model?"'.[6]

The JWST has already provided a useful check on Cepheid period–luminosity relations in Type Ia supernova host NGC 1365, the Great Barred Spiral Galaxy in the constellation Fornax.[7] NIRCam observations of more than 320 Cepheids in NGC 4258 and 5584 by Riess and his colleagues showed a factor of more than two improvement in the scatter of the period–luminosity relation. The results also showed no significant discrepancies with previous HST observations and so ruled out bias in the latter arising from confusion and crowding, providing 'the strongest evidence yet that systematic errors in HST Cepheid photometry do not play a significant role in the present Hubble Tension'.[8] In contrast, studies by Freedman and Madore of early JWST data on Cepheid variables in the galaxy NGC 7250 in the constellation Lacerta 'demonstrate that crowding/blending effects are a significant issue in a galaxy as close as 20 Mpc'.[9]

A status report on the CCHP posted to the preprint archive in August 2024 reported JWST NIRCam measurements based on three different standard candles, TRGB stars, J-Region Asymptotic Giant Branch (JAGB) stars, and Cepheids. The JAGB stars form a well-defined population of extremely red carbon stars of intermediate age, about 1 magnitude brighter in the near infrared compared with TRGB stars (and therefore capable of probing greater distances). The JAGB stars are relatively ubiquitous and easily identified, without the need for spectroscopy. Whilst their analysis is not as straightforward as the TRGB stars, they provide relief from some of the complications of the Cepheids.

The study focused on 10 nearby galaxies, each hosting Type Ia supernovae, including NGC 4258, whose distance was used as an anchor to calibrate the zero points for all three standard candles.

The resulting values for H_0 (in km/s/Mpc) are: 69.85 ± 2.33 (TRGB), 67.96 ± 2.65 (JAGB), and 72.05 ± 3.62 (Cepheids).* The CCHP team concluded: 'These differences are pointing to systematics

---

* These figures represent total errors, calculated as the square root of the (square of the reported statistical error plus the square of the reported systematic error).

affecting one or more of the distances and need to be better understood. However, while they do not rule it out, the results presented here do not lend strong support to the suggestion that there is missing fundamental physics in the early universe. Only future data will settle this issue unambiguously'.[10]

After nearly a century of measurements of the Hubble constant, we may yet be in the endgame. This is a rapidly evolving field and the interested reader is recommended to watch this space. Closely.

The Case for New Physics

Irrespective of how and when the secondary Hubble tension will be resolved to the satisfaction of all concerned, and irrespective of what this means for the primary tension, we might be willing to accept that it is important to start thinking now about what the primary tension between early-universe predictions and late-universe measurements might mean. Alternatively, we might just accept that theorists will be theorists. The merest hint that some form of 'new physics' might be required is far too tempting to resist, and conceiving new ideas and speculating on the basis of limited and uncertain data is what they do.

Any attempt to move beyond the standard Λ-CDM cosmology will almost by definition involve the introduction of new physics. Here 'new' does not imply a wholesale abandonment of the physical principles on which Λ-CDM cosmology is founded. The primary tension notwithstanding, it can be argued that the degree of concordance across a broad range of observational data means that it is premature to think about throwing this baby out with the bathwater.

As the early-universe predictions depend on the assumption of a specific cosmological model, it is here that the attentions of theorists tend to be focused. The late-universe distance ladder measurements rely on physical principles that are both well established and broadly unquestioned—all the quibbling is concerned with the observations themselves. But a complex theoretical model with many inter-related parameters will always be susceptible to tweaking. The aim of most of the tweaks attempted thus far has been to raise the early-universe predictions for H_0 to match the late-universe measurements. And, as the expansion rate is intimately linked to the density of dark energy, this is the knob that theorists have spent most time turning.

The theoretical solutions can be categorized as 'early time' and 'late time'.[11] The late-time solutions seek to raise the early-universe predictions for H_0 by *reducing* the density of dark energy following recombination compared with the Λ-CDM model, but keeping the energy density today (ρ_0) fixed. This means that the ratio ρ_z/ρ_0, where ρ_z is the energy density at redshift z, is lower than the corresponding ratio in the standard model. This can be achieved by postulating some form of exotic matter which contributes to dark energy but whose energy density *increases* with time. There are a number of physical objections to this proposal and, as it implies an equation of state parameter w less than -1, it is also potentially inconsistent with observation.

The best early-time solutions take the opposite approach. They postulate a no less exotic fluid that contributes about 10% to the density of dark energy but which decays to radiation rapidly following recombination, leaving us with only the vacuum energy of empty space. This larger Early Dark Energy (EDE) density at the time of recombination has the effect of shifting the sound horizon to larger angular scales (smaller l) but, as this is measured independently from baryon acoustic oscillations to be about 150 Mpc, the larger dark energy density must be compensated for by a larger H_0 so that the location of the sound horizon comes out right. To avoid charges of fine-tuning, attempts have been made to connect the EDE with other physical processes occurring around this time, so that any coincidence in timing is seen to arise naturally.

In March 2025, tantalizing evidence was presented for a cosmology featuring a dynamic dark energy which is weakening over time.[12] The Dark Energy Spectroscopic Instrument (DESI) collaboration based at Kitt Peak National Observatory in Arizona reported results from a BAO survey based on 14 million galaxies and quasars. Although the analysis yielded familiar values for some of the cosmological parameters $-H_0 = 68.45 \pm 0.47$ km/s/Mpc, and a sound horizon of 148 Mpc – a cosmological model featuring dynamic dark energy with an equation of state parameterised as $w(a) = w_0 + w_a(1 - a)$, with $w_0 > -1$ and $w_a < 0$ – was found to be favoured over Λ-CDM cosmology with a fixed dark energy density (i.e., a cosmological constant with $w = -1$).

In this expression the scale factor a is the reciprocal of $(1 + z)$, so for very high redshift (very small scale factors) the dark energy equation of state approaches the sum $w_0 + w_a$ and evolves over time to the present-day value w_0. Combining data from DESI and analysis of the cosmic background radiation yielded $w_0 = -0.42 \pm 0.21$ and $w_a = -1.75 \pm 0.58$, suggesting a decline in the strength or 'negative pressure' of dark energy from -2.17 at high redshift to -0.42 today. The DESI project is set to run for another four years, and it is too soon to draw firm conclusions (a fixed cosmological constant can't yet be ruled out), but here is yet another space to watch, closely.

There is a variety of Λ-CDM/EDE models, and no single observable consequence of the new physics beyond its impact on the expansion history of the universe. But there are nevertheless different possible observable consequences arising from different models, offering at least some vestige of hope that if the primary Hubble tension is indeed real, we might one day be able to explain its origin.

Other early-time solutions involve modifications to Einstein's general theory of relativity which serve to modify gravity at the time of recombination. Tweaking the physics of the primordial fluid just prior to recombination can help to shift the sound horizon without invoking EDE. These many different theoretical solutions have been reviewed, and ranked.[13] All we need now are more data.

I should also note in passing that as dark matter has such a pivotal role to play in the evolution of the universe, completing the Λ-CDM model to everybody's satisfaction will require us to understand what this is. If dark matter consists of particles of some kind, then be sure these are not to be found in the current standard model of particle physics. This is deeply embarrassing.

Could dark matter consist of primordial black holes, formed soon after the Big Bang? What about exotic astronomical objects composed of ordinary matter which emit little or no radiation (and which therefore appear 'dark'), called Massive Astrophysical Compact Halo Objects, or MACHOs? There's plenty of scope here for new physics. Sterile neutrinos? Axions? Supersymmetric particles? Weakly Interacting Massive Particles (WIMPs)? What if they are not particles at all? What if, instead of 'missing mass', we actually have an 'acceleration discrepancy'?[14]

One alternative that doesn't involve invoking mysterious new particles is to modify Newton's second law of motion, force = mass × acceleration, or $F = ma$. In this commonly accepted form, the law applies to stars in the outer regions of a spiral galaxy which experience a much reduced gravitational pull, such that they are expected to rotate around the centre of the galaxy much more slowly, in contradiction to the results of measurements of rotational speeds described in chapter 6. In Modified Newtonian Dynamics (MOND), developed in the early 1980s by Israeli physicist Mordehai Milgrom, the second law becomes $F = m(a^2/a_0)$, where a_0 is a new fundamental constant and, in the 'deep-MOND' limit, a is much smaller than a_0.

When plugged into Newton's gravitational law, the rotation speeds, v, of stars in the outer regions of a galaxy with total baryonic mass M can then be calculated as $v^4 = GMa_0$, where G is Newton's gravitational constant. This implies that the rotation speeds in the outer regions of a galaxy with mass M do not change as we look further and further away from the centre, consistent with measurements of the kind shown in Fig. 25. The plateaus in the rotation curves can be reproduced by assuming $a_0 = 1.2 \times 10^{-10}$ m/s².

As the theory draws on both the classical second law *and* the law of gravitation in combination, the same results can be obtained by modifying Newton's law of gravitation and leaving the second law intact. MOND offers the advantage that in situations where a is equal to or larger than a_0, the usual second-law behaviour (or Newton's gravity) is recovered. More sophisticated versions have been developed based on modifications of Einstein's general theory of relativity.

MOND and its successors have scored some notable successes on the scale of individual galaxies, but it cannot yet displace dark matter in all astrophysical systems where this is required to interpret observations. It doesn't describe the motions of galaxy clusters very well. It struggles to account for the temperature anisotropies in the cosmic background, and especially the amplitude of the third acoustic peak which, in models like MOND with no cold dark matter, is predicted to be lower than that of the second peak. On the other hand, the theory predicts that large-scale structures should form much more quickly than Λ-CDM can comfortably accommodate. This would seem to be entirely consistent with the evidence for such structures at very early times in the evolutionary history of the universe, as the JWST is now revealing.

For as long as searches for new particles with the properties required of dark matter continue to come up empty, the interest in various forms of modified dynamics, modified gravity, and modified general relativity will continue.

We are not short of ideas.

Rethinking Cosmology

There is, of course, another view. The Λ-CDM model relies heavily on several concepts for which, despite much effort over the last 20–30 years, we have secured absolutely no additional empirical evidence beyond the need for them. Cosmic inflation was introduced because it was needed to resolve a series of cosmological problems for which there were no ready solutions. Dark matter was introduced because it was needed to account for the formation and stability of galaxies and aspects of their large-scale motions. Dark energy was introduced because it was needed to account for the accelerating expansion of the universe. Note that I've chosen to use the words 'resolve' and 'account for' rather than 'explain', as their currently mysterious and unfathomable nature means that they actually don't *explain* anything at all.

In an open letter published 20 years ago in *New Scientist* magazine, a group of cosmologists declared: 'In no other field of physics would this continual recourse to new hypothetical objects be accepted as a way of bridging the gap between theory and observation. It would, at the least, raise serious questions about the validity of the underlying theory'.[15] Among the letter's 34 signatories were Hermann Bondi and Thomas Gold who, together with Hoyle, had pioneered steady-state cosmology in 1948. They had a point. If the Λ-CDM model is supposed to be the standard model of Big Bang cosmology, then we must accept that it falls short of the standards set by its counterpart from particle physics. Yes, the standard model of particle physics has problems of its own. It requires ~ 20 parameters which can only be obtained from experiment, but at least there is empirical evidence for all the particles that lie within its scope, which are all the particles that have been directly (and indirectly) observed.

Perhaps, after all, it is time to consider emptying the bath.

Steinhardt, one of the architects of inflationary cosmology in the early 1980s, has long been haunted by second thoughts. In the public announcement of the Planck 2013 results, these were lauded by the Planck team as a powerful confirmation of inflationary Λ-CDM cosmology. But, gathered at the Harvard–Smithsonian CfA a short time later, Steinhardt, on sabbatical at Harvard, his host Abraham Loeb, chairman of the astronomy department, and Anna Ijjas, a visiting graduate student, drew rather different conclusions.[16] Recall that the theorists gathered at the Nuffield workshop in 1982 had found that the simplest inflationary models predict much stronger temperature fluctuations in the cosmic background, inconsistent with the Planck results. Steinhardt and Turner had turned the problem on its head, using an inflaton field to drive inflation in a way that could be tuned to reproduce fluctuations of the right size. But this makes inflation theory extraordinarily flexible. Change the starting conditions and the properties of the inflaton field and the theory can be adjusted to fit the observed patterns, undermining at least one of the solutions—to the problem of fine-tuning—that inflation was supposed to provide.

It gets worse. Heisenberg's uncertainty principle quite happily accommodates fluctuations in a quantum field with substantial energies, provided these occur in a sufficiently short time that their product remains within its bounds. Substantial fluctuations in different regions of the inflaton field can in principle drive inflation much faster than the field decays, producing large volumes of exponentially expanding space which become dominant. Any area that exits the inflationary phase produces a bubble of hot matter and radiation which we recognize as 'the' Big Bang.

But such areas are surrounded by regions that continue to inflate, producing bubble after bubble after bubble.[17] The result of such 'eternal inflation' is a 'multiverse', an infinity of universes all with different, randomly selected cosmological properties.[18] This universe over here expands too rapidly to allow stars and galaxies to form. In this other universe over there space is negatively curved. In such a scenario, the only way to answer to the question; 'Why is our universe the way it is?' is to invoke something called the anthropic cosmological principle.[19] Our (Goldilocks) universe is the way it is because otherwise we wouldn't be here to observe it.

The anthropic principle is deeply divisive. Some theorists have embraced it as the answer to virtually all of the stubborn 'why' questions that we can't otherwise provide ready answers for. Other theorists (and most astronomers) dismiss it as circular, at best, or incoherent metaphysics at worst. Ijjas, Steinhardt, and Loeb (ISL) pulled no punches. 'We would like to suggest "multimess" as a more apt term ... A good scientific theory is supposed to explain why what we observe happens instead of something else. The multimess fails this fundamental test'. They argued that inflation lies at the root of the problem. It is simply not susceptible to empirical test: '... these features [the subjective choice of starting conditions and properties of the inflaton field] make inflation so flexible that no experiment can ever disprove it'. It is, as philosophers of science say, irrefutable. Consequently, ISL argue, inflation theory is not scientific.

Their solution? Abandon Big Bang Λ-CDM cosmology in favour of a bouncing model which does not require an inflationary phase.

Not everybody agreed. The ISL article published in *Scientific American* in 2017 attracted a robust response from no less than 33 cosmologists, including Guth, Linde, Hawking, Krauss, Mather, Rees, Smoot, Turner, Weinberg, and Vilenkin.[20] Weinberg has claimed that he does not believe physicists will become so comfortable with anthropic reasoning that they will give up the pursuit of first principles. 'I don't know of any theorist who would not prefer ... an alternative theory to anthropic reasoning, and who thinks we should give up the search for such a theory. It's just not an easy search, and for all we know we may be searching for something that does not exist'.[21] The ISL position remains controversial.

If we care to look, there are plenty more problems to be found in addition to the primary Hubble tension (if it is real), the trouble brewing from the early galaxies found by the JWST, and the irrefutability of inflation.[22] Several of these are inflation-related. Others are more foundational. There are longstanding problems with the interpretation of the quantum theory on which the early stages of Λ-CDM cosmology depends, especially with regard to the transition between small-scale quantum behaviour and large-scale classical behaviour of the kind described by the general theory of relativity. These problems could perhaps be resolved by combining quantum mechanics and general relativity in a quantum theory of gravity. But, although theorists have been working hard for many years to find this theory, there is no consensus even on which approach to take.[23]

Going right back to basics, there are inconsistencies in the way in which the Friedmann version of Einstein's gravitational field equations is employed. Assuming the cosmological principle—the universe is homogeneous and isotropic—is arguably questionable during phases of accelerated expansion. There are echoes here of the original arguments in favour of steady-state cosmology.

A recent experiment at CERN in Geneva demonstrated that antimatter, in the form of neutral anti-hydrogen atoms (consisting of an antiproton and positron) fall *downwards* in Earth's gravitational field, just like ordinary matter.[24] This was unsurprising to many, but leaves a big headache for a Big Bang cosmology which would be expected to produce equal amounts of matter and antimatter in the primeval universe (equal numbers of electrons, e^-, and positrons, e^+, for example). When matter particles encounter antimatter particles, they annihilate to produce energetic photons.

So, we're left to ponder why the universe has any matter in it at all or, alternatively, where all the antimatter has gone. This problem of *baryon asymmetry* has no solution, but it was thought that if matter and antimatter experienced a repulsive form of gravity, then instead of being drawn together to their doom, all the matter and antimatter might have rather been flung apart and separated to form a universe, and an anti-universe. We can now rule out this idea, leaving the problem unresolved.

Finally, there remains the 'mass gap' problem of Big Bang nucleosynthesis. Whilst it correctly predicts the observed abundances of primordial hydrogen and helium, it predicts three-to-four times more lithium than we can find. I should acknowledge that not all of these challenges are of equal import. There are physically reasonable explanations for some of them, though they carry insufficient consensus within the community for the challenges to be considered resolved.

Future Experience

Take heart. The arguments in favour of new physics and the demands to rethink cosmology should not detract from the extraordinary achievements of astronomy and cosmology over the past century or so. What we are witnessing is simply the scientific enterprise at work, and this is often messy and incoherent. Answers to some of our deepest questions about the universe and our place in it can sometimes appear frustratingly incomplete. There is no denying that, for all its faults, the

Λ-CDM model now dominates the science of cosmology, for good reasons. But the lessons from the troubled history of the Hubble constant warn against becoming too comfortable. There is undoubtedly more to discover about our universe. There will be more surprises.

The challenges are, as always, to retain a strong sense of humility in the face of an inscrutable universe, and to keep an open mind. As Einstein put it: 'The *truth* of a theory can never be proven, for one never knows if future experience will contradict its conclusions'.[25]

Astronomy is an expensive science, and throughout its history it has relied heavily on support from both wealthy benefactors and taxpayers. Financing big, billion-dollar scientific programmes is always fraught with a degree of uncertainty, and it pays scientists never to take funding for granted. But within the last year, American astronomers have been confronted with an unprecedented attack by their own government on academic institutions and science funding. At the time of correcting the proofs of this book (June 2025), the outlook for American science was bleak, with Riess declaring that proposed deep cuts to NASA's budget – the largest single-year decline in its history – are " . . . almost extinction level for a lot of the space science that we pursue". Another space to watch closely, though this with some trepidation.[26]

Symbols and Acronyms

FREQUENTLY USED SYMBOLS

a_t	Scale factor at time t
a_0	Scale factor at time t_0 (now)
c	Speed of light in a vacuum
d	Distance (also proper distance)
d_t	Distance (e.g., between galaxies) at time t
d_0	Distance (e.g., between galaxies) at time t_0 (now)
G	Newton's gravitational constant
h	Dimensionless Hubble constant = $H_0/(100 \text{ km/s/Mpc})$
H_0	Hubble constant at time t_0 (now)
K	Kelvin (temperature scale: 0 K = −273.15 Celsius)
Λ	Cosmological constant
l	Multipole moment (TT power spectrum)
ly	Light-year (unit of distance)
L	Length or distance parameter = a_t/a_0
m	Apparent magnitude
m_p	Photographic magnitude
m_v	Visual magnitude
m_T	Total apparent magnitude
M	Absolute magnitude
Mpc	Megaparsec (= 1 million parsecs)
μ	Distance modulus = $m - M$
n_s	Spectral index (cosmological parameter)
Ω	Density parameter = $\rho/\rho_c = \Omega_M + \Omega_\Lambda$
Ω_B	Density parameter for baryonic matter = ρ_B/ρ_c
Ω_{DM}	Density parameter for dark matter = ρ_{DM}/ρ_c

Continued

Continued

Ω_M	Density parameter for matter $= \Omega_B + \Omega_{DM}$
Ω_Λ	Density parameter for dark energy $= \rho_\Lambda / \rho_c$
P	Period (of Cepheid variables, measured in days)
pc	Parsec (unit of distance)
ρ_B	Density of baryonic matter
ρ_c	Critical density of mass–energy required for a spatially flat universe
ρ_{DM}	Density of dark matter
ρ_Λ	Density of dark energy
ρ_M	Density of matter $= \rho_B + \rho_{DM}$
ρ_r	Density of radiation
q_0	Deceleration parameter at time t_0 (now)
T	Temperature (in kelvin)
τ	Reionization optical depth (cosmological parameter)
θ	Angular scale (TT power spectrum) $\sim 180°/l$
v	Radial (line-of-sight) recession speed of galaxies
w	Dark energy equation of state (cosmological parameter)
z	Redshift $= \sqrt{(1 + v/c)/(1 - v/c)} - 1$. If v is very much smaller than c, this can be approximated as $z \cong v/c$

Acronyms

2dFGRS	Two-degree Field Galaxy Redshift Survey
ACT	Atacama Cosmology Telescope
ARC	Astrophysical Research Consortium
BOOMERANG	Balloon Observations Of Millimetric Extragalactic Radiation and Geophysics (project)
BOSS	Baryon Oscillation Spectroscopic Survey (project)
CATS	Comparative Analysis of TRGBs
CCD	Charge-coupled device (image sensor)
CCHP	Carnegie–Chicago Hubble Program
CDM	Cold Dark Matter
CEERS	Cosmic Evolution Early Release Science Survey (project)
CfA	Harvard–Smithsonian Center for Astrophysics
CfPA	Center for Particle Astrophysics (LBL/UC Berkeley)
COBE	Cosmic Background Explorer (satellite mission)
COSMOGRAIL	Cosmological Monitoring of Gravitational Lenses (project)

DESI	Dark Energy Spectroscopic Instrument
DMR	Differential Microwave Radiometer (instrument aboard COBE)
EDE	Early Dark Energy (extensions of Λ-CDM model)
EDR3	Early Data Release 3 (Gaia Space Observatory)
ESA	European Space Agency
EVA	Extra-vehicular Activity (space walk)
FIRAS	Far InfraRed Absolute Spectrophotometer (instrument aboard COBE)
FLRW	Friedman–Lemaître–Robertson–Walker metric
Gaia	Global Astrometric Interferometer for Astrophysics (space observatory)
H_0LiCOW	H_0 Lenses in COSMOGRAIL's Wellspring (project)
HAPPE	High Altitude Particle Physics Experiment (project)
H-R	Hertzsprung–Russell (diagram)
HST	Hubble Space Telescope
IAU	International Astronomical Union
IRAS	Infrared Astronomical Telescope
JADES	JWST Advanced Deep Extragalactic Survey (project)
JAGB	J-Region Asymptotic Giant Branch
JWST	James Webb Space Telescope
Laser	Light amplification by the stimulated emission of radiation
LBL (LBNL)	Lawrence Berkeley Laboratory (Lawrence Berkeley National Laboratory)
LRG	Luminous Red Galaxy
Λ-CDM	Lambda-Cold Dark Matter model (cosmology)
MAP	Microwave Anisotropy Probe (satellite mission, renamed WMAP)
Maser	Microwave amplification by the stimulated emission of radiation
MIT	Massachusetts Institute of Technology
MOND	Modified Newtonian Dynamics (theory)
NASA	National Aeronautics and Space Administration (US)
NICMOS	Near Infrared Camera and Multi-object Spectrometer (instrument aboard the HST)
NIRCam	Near-Infrared Camera (instrument aboard the JWST)
NIRSpec	Near-Infrared Spectrograph (instrument aboard the JWST)
NGC	New General Catalogue (of nebulae and clusters of stars)
PL	Period–luminosity (relation)
PLC	Period–luminosity–colour (relation)
PSCz	IRAS Point Source Catalog Redshift (survey)
RAND	Research and Development (non-profit organization)
SCP	Supernova Cosmology Project
SDSS	Sloan Digital Sky Survey
SH_0ES	Supernovae and H_0 for the Equation of State of Dark Energy (project)

Continued

Continued

SN/SNe	Supernova/supernovae
SPT	South Pole Telescope
SQL	Structured Query Language
STRIDES	Strong Lensing Insights into Dark Energy Survey (project)
STScI	Space Telescope Science Institute
TRGB	Tip of the Red Giant Branch (stars)
WF/PC	Wide Field and Planetary Camera (instruments aboard the HST)
WFC3	Wide Field Camera 3 (instrument aboard the HST)
WMAP	Wilkinson Microwave Anisotropy Probe (satellite mission)

Cosmic Distances

'Space', [the Hitch-Hiker's Guide to the Galaxy] says, 'is big. Really big. You just won't believe how vastly, hugely, mindbogglingly big it is. I mean you may think it's a long way down the road to the chemist, but that's just peanuts to space'.

Douglas Adams, *The Hitch-Hiker's Guide to the Galaxy*

Understandably, astronomers cannot afford to spend too much of their time being mindboggled by the vast distances of the objects they observe in the universe. They long ago adopted systems of units much more suited to such distances than terrestrial miles or kilometres, of the kind we might use to determine the distance to the local chemist.

In 1976, the IAU adopted the **astronomical unit**, used routinely by astronomers for reporting distances within the solar system. The symbol 'au' was adopted in 2012. The au is based on the distance between the Sun and the Earth, about 150 million km (or, more precisely, 1 au = 149,597,870,700 metres).

Perhaps the most well-known astronomical distance measure is the **light-year** (ly)—the distance that light travels through space in a year. This can be calculated simply as the speed of light in a vacuum multiplied by the time that elapses in a year. The IAU uses a Julian year (365.25 days), or 31,557,600 seconds. With the speed of light given as exactly 299,792,458 m/s, we can define a light-year to be equivalent to 9.461 trillion (million million) km. Our nearest stellar neighbour, Proxima Centauri, lies about 4.25 light-years away.

The light-year and astronomical unit are useful distance units but these are not based on any means by which the distances to distant objects are actually measured. For relatively nearby objects, distances can be measured from the apparent shift in position in the sky when viewed along two different lines of sight against a fixed background. The range of distance that can be measured depends on the length of the baseline of the triangle that can be drawn to a distant point, and hence the sharpness of the angle at its apex (see Fig. 3). The 'parallax angle' is half of this.

The **parsec** (parallax of one second, pc) is then a measure of distance corresponding to a parallax angle of 1 arcsecond, or 1/60 of an arcminute, or 1/3600 of a degree, subtended by the Sun–Earth distance of 1 au (corresponding to half the baseline of the triangle). If the parallax angle is small, then the distance in parsecs can be approximated as the reciprocal of the angle measured in arcseconds. Proxima Centauri has a parallax angle of 0.769 arcseconds, corresponding to a distance of 1.30 pc. A **megaparsec** (Mpc) is a million pc.

How these various distance measures relate to each other is shown in Table A1.

Table A1. Comparing cosmic distances

	Equals ...	Equals ...	Equals ...	Equals ...
1 au	1 au	4.848×10^{-6} ly	1.581×10^{-5} pc	149,597,871 km
1 ly	63,241 au	1 ly	0.3066 pc	9.461×10^{12} km
1 pc	206,265 au	3.262 ly	1 pc	3.086×10^{13} km
1 Mpc	206×10^9 au	3.262×10^6 ly	10^6 pc	3.086×10^{19} km

Distance measures based on standard candles such as Cepheid variables or TRGB stars make use of the system of units defined above. Larger distances are determined from measures of the redshift of spectral lines using the Hubble–Lemaître law which, for instances where the redshift can be approximated as $z \sim v/c$, can be written as $cz = H_0 d$, or $d = cz/H_0$. If the units of c are km/s and the units of H_0 are km/s/Mpc, then the distances d are measured in Mpc.

Redshifts and 'Look-back' Times

It is not possible to give a simple expression for the relation between redshift z and its associated 'look-back' time. Their relationship depends in a complex way on the choice of a specific cosmology and a set of cosmological parameters.

However, there are a number of 'cosmology calculators' available online, and a small selection is published here: https://ned.ipac.caltech.edu/help/cosmology_calc.html.

For many years I have used Ned Wright's cosmology calculator: https://www.astro.ucla.edu/~wright/CosmoCalc.html. This allows the user to select values for H_0, Ω_M, and (by choosing a spatially flat universe), $\Omega_\Lambda = 1 - \Omega_M$. For any input value for z, the calculator returns the age of the universe at the time the light was emitted (the age at redshift z), the light travel time (the 'look-back' time), and the comoving radial distance d in Mpc and light-years.

For the calculations shown in Fig. A1 below, I set $H_0 = 70$ km/s/Mpc, $\Omega_M = 0.3$, and $\Omega_\Lambda = 0.7$, corresponding to a universe with an age of 13.462 billion years.

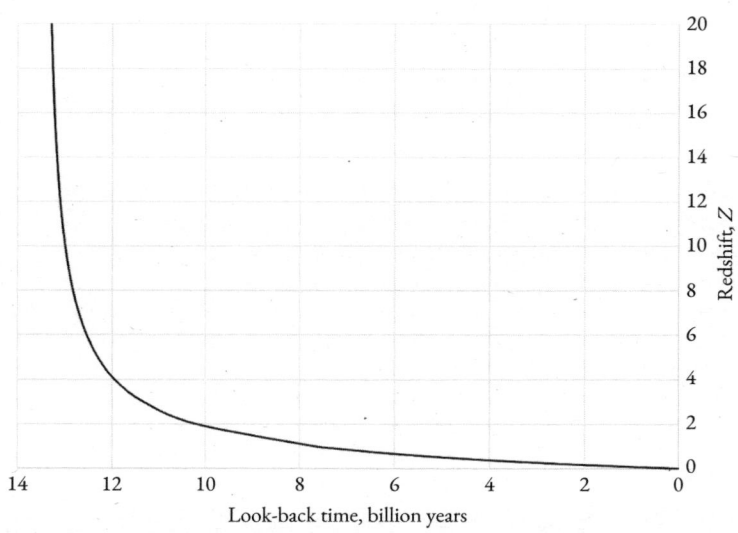

Fig. A1 The redshift of distant objects is an indicator of distance according to $d = cz/H_0$. But as it takes a finite time for the emitted light to reach us, the redshift also tells us about objects as they appeared when the light was emitted. This is the 'look-back' time. A redshift $z = 20$ corresponds to a look-back time of 13.287 billion years, 98.7% of the age of the universe assumed in these calculations.

Acknowledgements

I think it's fair to say that, like the eight-year-old Edwin Hubble, most kids are fascinated by outer space and space exploration. You don't need to take my word for it. In 2019, on the 50th anniversary of the Apollo 11 Moon landing, the toy company LEGO commissioned a poll among children aged 8–12 in the US, UK, and China. This was conducted by The Harris Poll.[*] When asked if they are interested in space exploration, an extraordinary 86% of those surveyed responded positively, and 90% said they wanted to learn more. As many as 85% identified Neil Armstrong as the first astronaut to walk on the Moon (and only 2% thought he was Buzz Lightyear).

For all but a few, this enthusiasm fades with age and maturity, as more earthly concerns overtake and dominate thought. Childish ideas are set aside in favour of more practical pursuits. Those few whose interest to learn more sustains an early exposure to the sciences of astronomy, astrophysics, and cosmology face more potential hurdles. Interest can be all-too-easily broken on the rocks of physics, and its attendant Grim Reaper of youthful enthusiasm, mathematics. For many, childhood fascination with space never develops beyond a vague dream of one day travelling to Mars. The translation of such fascination into a Hubble-esque commitment is a rarity.

I hope that *Discordance* helps to keep at least some dreams alive. In telling the story of how we come to know what we do about our universe, my challenge has been to expose enough of the physics and mathematics for readers to get a sense for how the story has developed over the past 120 years or so. The science can be hard, but the reward for sticking with it is the glimpse it affords of the extraordinary beauty of the universe. Not just the universe as we see it, but the universe as we try to comprehend it. Our current description also offers a deeper appreciation of its inscrutability. There's a lot we don't know. But we still know quite a lot.

I could not have risen to this challenge alone. I've received support and encouragement from Wendy Freedman at the University of Chicago and Adam Riess at the STScI, for which I'm extremely grateful. I also thank George Ellis at the University of Cape Town and Joe Silk at Oxford University for their thorough and insightful reviews of the manuscript. For a few years now, I've subscribed to *A Cosmology Group*, a bunch of renegade cosmologists whose email exchanges and newsletters have provided useful notifications of contemporary developments.

I have been privileged to have had eight books edited and guided to publication by Latha Menon at Oxford University Press. That this was our last collaboration is a source of great sadness, tempered somewhat by the opportunity it afforded to return Latha to an early

[*] https://theharrispoll.com/briefs/lego-group-kicks-off-global-program-to-inspire-the-next-generation-of-space-explorers-as-nasa-celebrates-50-years-of-moon-landing/.

love, astronomy and astrophysics. I am also grateful to Molly Balikov, Jamie Mortimer, and the production team at OUP. Their ability to turn a manuscript and a collection of diagrams and photographs into a handsome volume is little short of astonishing, bordering on miraculous.

Jim Baggott
June 2025

Figure & Photo Credits

Endnotes

PROLOGUE: STARRY MESSENGERS

1. https://www.constellation-guide.com/orions-belt/.
2. See Michael Rappenglück, 'The Anthropoid in the Sky: Does a 32000-year-old Ivory Plate Show the Constellation Orion Combined with a Pregnancy Calendar?', *Proceedings of the 9th Annual Meeting of the European Society for Astronomy in Culture (SEAC)*, Stockholm, 27–30 August 2001, Uppsala Astronomical Observatory Report No. 59 (2003).
3. http://www.atlasoftheuniverse.com/stars.html.
4. Norman Pogson, 'Magnitudes of Thirty-six of the Minor Planets for the First Day of Each Month of the Year 1857', *Monthly Notices of the Royal Astronomical Society*, **12** (1856) 14.
5. Gustav Theodor Fechner, *Elemente der Psychophysik*, Breitkopf und Härtel, Leipzig, 1860. The first volume of this two-volume work has been translated into English as *Elements of Psychophysics*, Volume I, Holt, Rinehart and Winston, New York, 1966.
6. I. Newton, *Philosophical Transactions of the Royal Society*, **6** (1672) 3075.

CHAPTER 1: MISS LEAVITT'S LAW

1. Bessie Zaban Jones and Lyle Gifford Boyd, *The Harvard College Observatory: The First Four Directorships 1839–1919*, Harvard University Press, 1971, p. 213.
2. 'The Arequipa Observatory', *Nature*, **57** (1898) 249–251.
3. Edward Pickering, quoted in George Johnson, *Miss Leavitt's Stars*, W.W. Norton & Company, New York, 2005, p. 18
4. Antonia C. Maury, 'Spectra of Bright Stars', *Annals of the Astronomical Observatory of Harvard College*, XXVIII-Part I (1897), p. 5.
5. H.N. Russell, 'Relations Between the Spectra and Other Characteristics of the Stars', *Popular Astronomy*, **22** (1914) 275–294. This quote appears on p. 276.
6. E. Hertzsprung, letter to E.C. Pickering, 22 July 1908. Quoted in Jones and Boyd, p. 240.
7. S.I. Bailey, 'Henrietta Swan Leavitt', *Popular Astronomy*, **30** (1922) 197–199. This is Leavitt's obituary—she died of cancer in December 1921.
8. H.S. Leavitt, '1777 Variables in the Magellanic Clouds', *Annals of the Astronomical Observatory of Harvard College*, **60** (1908) 107.
9. Henrietta Leavitt, 'Periods of 25 Variable Stars in the Small Magellanic Cloud', *Harvard College Observatory Circular*, No. 173 (3 March 1912).
10. A.S. Eddington, *Stellar Movements and the Structure of the Universe*, MacMillan & Co., London, 1914, p. 171.

11. E. Hertzsprung, 'Über die Räumliche Verteilung der Veränderlichen vom δ Cephei-Typus', *Astronomische Nachrichten*, **196** (1913), 201–208. See also J.D. Fernie, 'The Period–Luminosity Relation: A Historical Review', *Publications of the Astronomical Society of the Pacific*, **81** (1969), 707–731.

CHAPTER 2: THE SCALE OF THE UNIVERSE

1. William Wordsworth, 'The stars are mansions built by Nature's hand'.
2. https://astrobiology.nasa.gov/news/water-on-mars-the-story-so-far/.
3. See Kevin S. Schindler, 'Through the Lens of History: The Unusual Circumstances Leading to the Acquisition of the Lowell Spectrograph', in Michael J. Way and Deidre Hunter, eds, *Origins of the Expanding Universe: 1912–1932*, Astronomical Society of the Pacific Conference Series, **471** (2013) 174–175.
4. W.G. Hoyt, 'Vesto Melvin Slipher, 1875–1969', *Biographical Memoirs*, National Academy of Sciences, **42** (1980) 422.
5. V.M. Slipher, 'The Radial Velocity of the Andromeda Nebula', *Lowell Observatory Bulletin*, **2** (1913) 56.
6. Percival Lowell, letter to Vesto Slipher, 8 February 1913, Lowell Observatory Archives. Quoted by Hoyt, p. 424.
7. V.M. Slipher, 'Spectrographic Observations of Nebulae', Abstract, 17th AAS Meeting, Evanston, IL, published in *Popular Astronomy*, **23** (1915) 21–24. This quote appears on p. 23.
8. See Helge S. Kragh, *Conceptions of Cosmos*, Oxford University Press, 2007, p. 117.
9. V.M. Slipher, 'Nebulae', *Proceedings of the American Philosophical Society*, **56** (1913) 409.
10. Albert Einstein, letter to Paul Ehrenfest, 4 February 1917, quoted in Abraham Pais, *Subtle is the Lord: The Science and the Life of Albert Einstein*, Oxford University Press, 1982. The quote appears on p. 285.
11. A.S. Eddington, *The Expanding Universe*, Cambridge University Press, 1933, p. 21.
12. Newton wrote: '... and lest the systems of the fixed stars should, by their gravity, fall on each other mutually, he [God] hath placed those systems at immense distances one from another'. Isaac Newton, *Mathematical Principles of Natural Philosophy*, first American edition translated by Andrew Motte, published by Daniel Adee, New York, 1845, p. 504.
13. A. Einstein, 'Cosmological Considerations in the General Theory of Relativity', *Proceedings of the Prussian Academy of Sciences*, **142** (1917). Quoted in Walter Isaacson, *Einstein: His Life and Universe*, Simon & Shuster, New York, 2007, p. 255.
14. H.N. Russell, 'Notes on the Real Brightness of Variable Stars', *Science*, **37** (1913) 651.
15. Harlow Shapley, *Through Rugged Ways to the Stars: The Reminiscences of an Astronomer*, Charles Scribner's Sons, New York, 1969, p. 17.
16. Shapley, *Through Rugged Ways*, p. 44.
17. H. Shapley, 'Studies Based on the Colors and Magnitudes in Stellar Clusters, Sixth Paper: On the Determination of the Distances of the Globular Clusters', *Astrophysical Journal*, **18** (1918) 89.
18. For example, van Maanen determined the rotation period of the spiral nebula M101 (now known as the Pinwheel galaxy) to be of the order of 85,000 years (or 2.7×10^{12} s). If we assume this to be a galaxy of the same order of size as Shapley's Milky Way, this puts the diameter of M101 at 300,000 light-years, or a radius of 150,000 light-years

$(1.4 \times 10^{18}$ km$)$. If we assume it to rotate as a rigid body, its rotation speed is then given by $2\pi \times 1.4 \times 10^{18}/2.7 \times 10^{12}$ km/s, or 3.3 million km/s. The speed of light is about 300,000 km/s.

19. Ron Voller, *Hubble, Humason, and the Big Bang: The Race to Uncover the Expanding Universe*, Springer Praxis, Cham, 2021, p. 119.

20. Gale E. Christianson, *Edwin Hubble: Mariner of the Nebulae*, Institute of Physics Publishing, Bristol, 1997, p. 65.

21. Interview of Harlow Shapley by Charles Weiner and Helen Wright on 8 June 1966, Niels Bohr Library & Archives, American Institute of Physics, College Park, MD USA, p. 69.

22. Paul W. Merrill, Memories of George E. Hale and Mt Wilson, 1953, Niels Bohr Library & Archives, American Institute of Physics. Quoted by Voller, p. 195.

23. Christianson, p. 151.

24. Cecilia Payne-Gaposchkin, *An Autobiography and Other Recollections*, Katherine Haramundanis, ed., Cambridge University Press, 1984, p. 209. Quoted by Christianson, p. 159.

25. Interview of Harlow Shapley, p. 84.

26. Edwin Hubble, *The Realm of the Nebulae*, Yale University Press, 1936, p. li.

CHAPTER 3: HUBBLE'S CONSTANT

1. Ernst Mach, *The Science of Mechanics*, Open Court Press, Chicago, Il., 1893, pp. 224, 229 and 232.

2. Albert Einstein, *Relativity: The Special and the General Theory*, 100th Anniversary Edition, Princeton University Press, 2015, pp. 163–4.

3. See Isaac Newton, *Mathematical Principles of Natural Philosophy*, first American edition translated by Andrew Motte, published by Daniel Adee, New York, 1845, p. 73. In his Definition III, Newton writes: 'The *vis insita*, or innate force of matter, is a power of resisting, by which every body, as much as in it lies, endeavours to persevere in its present state, whether it be of rest, or of moving uniformly forward in a right [straight] line'.

4. Ernst Mach, *The Science of Mechanics*, Open Court Press, Chicago, Il., 1893, p. 232.

5. A. Einstein, 'On the Foundations of the General Theory of Relativity', *Annalen der Physik*, **55** (1918) 241–244.

6. W. de Sitter, 'On Einstein's Theory of Gravitation, and its Astronomical Consequences', *Monthly Notices of the Royal Astronomical Society*, **78** (1917) 3–28. This quote appears on p. 6.

7. W. de Sitter, pp. 26 and 28.

8. Alexander A. Friedmann, *The World as Space and Time*, translated by Svetla Kirilova-Petkova and Vesselin Petkov, edited by Vesselin Petkov, Minkowski Institute Press, Montreal, 2014, p. 80.

9. A. Einstein, 'Bemerkung zu der Arbeit von A. Friedman "Über die Krümmung des Raumes"', *Zeitschrift für Physik*, **11** (1922) 326.

10. A. Einstein, 'Notiz zu der Arbeit von A. Friedman "Über die Krümmung des Raumes"', *Zeitschrift für Physik*, 16 (1923) 226. The original handwritten manuscript of this note is in the Einstein Archives. It is reproduced in Harry Nussbaumer and Lydia Bieri, *Discovering the Expanding Universe*, Cambridge University Press, 2009, p. 91.

11. A.S. Eddington, *The Mathematical Theory of Relativity*, Cambridge University Press, 1923, p. 162.

12. G. Lemaître, 'Note on de Sitter's Universe', *Journal of Mathematics and Physics*, **4** (1925) 188–192. These quotes appear on p. 192.

13. The full expression for the redshift parameter is $z = \sqrt{(1 + v/c)/(1 - v/c)} - 1$, where v is the radial (line-of-sight) velocity of the object and c is the speed of light. But in situations where v is very much smaller than c, this can be approximated as $z \cong v/c$.

14. G. Lemaître, 'Rencontres avec A. Einstein', *Revue des Questions Scientifiques*, **129** (1958) 129–132. These quotes appear (in French) on p. 129. This is a text read on Belgian national radio on 27 April 1957 in commemoration of the second anniversary of Einstein's death.

15. E. Hubble, 'A Relation Between Distance and Radial Velocity Among Extra-Galactic Nebulae', *Proceedings of the National Academy of Sciences*, **15** (1929) 168–173. This quote appears on p. 173.

16. M.L. Humason, 'The Large Radial Velocity of N.G.C. 7619', *Proceedings of the National Academy of Sciences*, **15** (1929) 167–168. This quote appears on p. 167.

17. Edwin Hubble, letter to Willem de Sitter, 23 September 1931. Quoted in Robert W. Smith, *The Expanding Universe: Astronomy's 'Great Debate' 1900–1931*, Cambridge University Press, 1982, p. 192.

18. See Harry Nussbaumer and Lydia Bieri, *Discovering the Expanding Universe*, Cambridge University Press, 2009, p. 120.

19. Nussbaumer and Bieri, p. 121.

20. G.C. McVittie, 'Georges Lemaître', *Quarterly Journal of the Royal Astronomical Society*, **8** (1967) 294–297. McVittie's recollection of Eddington's embarrassment appears on p. 295.

21. M. Livio, 'Mystery of the Missing Text Solved', *Nature*, **479** (2011) 171–173. This quote from Lemaître's letter to astronomer William Marshall Smart, editor of the *Monthly Notices*, appears on p. 173.

22. Helge Kragh, 'Hubble Law or Hubble-Lemaître Law? The IAU Resolution', arXiv [physics-hist-ph] 1809.02557v1, 7 September 2018.

23. 'Redshift of nebulae a puzzle, says Einstein', New York Times, 12 February 1931, p. 2. Quoted in C. O'Raifeartaigh and B. McCann, 'Einstein's Cosmic Model of 1931 Revisited: An Analysis and Translation of a Forgotten Model of the Universe', *The European Physical Journal H*, February 2014, 1–23. This quote appears on p. 1.

24. A. Einstein, 'Zum Kosmologischen Problem der Allgemeinen Relativitätstheorie', *Sitzungsberichte der Preussischen Akademie der Wissenschaften, Physikalisch-mathematische Klasse*, (1931) 257–265. Quoted in English translation in C. O'Raifeartaigh and B. McCann. This quote appears on p. 4.

25. George Gamow wrote: 'When I was discussing cosmological problems with Einstein he remarked that the introduction of the cosmological term was the biggest blunder he ever made in his life'. George Gamow, *My World Line: An Informal Autobiography*, Viking Press, New York, 1970, p. 149. Quoted in Isaacson, pp. 355–356.

26. A. Einstein and W. de Sitter, 'On the Relation Between the Expansion and the Mean Density of the Universe', *Proceedings of the National Academy of Sciences*, **18** (1932) 213–214.

27. A. Knopf, C. Schuchert, A.F. Kovarik, A. Holmes, and E.W. Brown, 'Physics of the Earth—IV: The Age of the Earth', *Bulletin of the National Research Council of the National Academy of Sciences*, **80** (1931). Quoted in Cherry L.E. Lewis, 'Arthur Holmes' Unifying Theory: From Radioactivity to Continental Drift', *Geological Society, London, Special Publications*, **192** (2002) 167–183. The quote appears on p. 178.

CHAPTER 4: DIVINE CURVES OF CREATION

1. A.S. Eddington, *The Expanding Universe*, Cambridge University Press, 1933. On p. 24 Eddington writes: '... return to the earlier view is unthinkable. I would as soon think of reverting to Newtonian theory as of dropping the [cosmological] constant'.
2. Eddington *The Expanding Universe*, 'Views as to the beginning of things', p. 55; 'Since I cannot avoid introducing this question', p. 56, italics in the original; 'Have it your own way', p. 60.
3. G. Lemaître, 'The Beginning of the World from the Point of View of Quantum Theory', *Nature*, **127** (1931) 447.
4. This manuscript is held in the Georges Lemaître Archives, UC Louvain. Quoted by Helge S. Kragh, *Conceptions of Cosmos*, Oxford University Press, 2007, p. 175.
5. G. Lemaître, 'L' Expansion de l'Espace', *Revue des Questions Scientifiques*, **17** (1931) 391–410. Quoted in English translation in Kragh, *Conceptions of Cosmos*. This quote appears on p. 154.
6. G. Lemaître, 'Evolution of the Expanding Universe', *Proceedings of the National Academy of Sciences*, **20** (1934) 12–17. This is an address that Lemaître delivered to the Academy in November 1933. This quote appears on p. 12.
7. In 1933, Canadian astronomer John Plaskett declared Lemaître's fireworks theory to be 'the wildest speculation', nothing less than 'an example of speculation run mad without a shred of evidence to support it'. Quoted in Kragh, *Conceptions of Cosmos*, p. 157.
8. Edwin Hubble, *The Realm of the Nebulae*, Yale University Press, 1936, p. 202.
9. A.S. Eddington, 'The Internal Constitution of the Stars', *Nature*, **106** (1920) 14–20. This is Eddington's opening address to the Mathematical and Physical Science section at a meeting of the British Association in Cardiff on 24 August 1920. This quote appears on p. 19.
10. Cecilia H. Payne, 'Stellar Atmospheres', *Harvard Observatory Monographs*, **1** (1925). This is Payne's PhD thesis. The quote appears on p. 186.
11. H.A. Bethe, 'Energy Production in Stars', *Physical Review*, **55** (1939) 434–456.
12. Joel Stebbins, quoted by Alan Sandage in his obituary of Baade, *Quarterly Journal of the Royal Astronomical Society*, **2** (1961) 118–121. The quote appears on p. 119.
13. W. Baade, 'The Resolution of Messier 32, NGC 205, and the Central Region of the Andromeda Nebula', *Astrophysical Journal*, **100** (1944) 137–146. 'This leads to the further conclusion', p. 137; 'Although the evidence presented', p. 145.
14. W. Baade and F. Zwicky, 'On Super-novae', *Proceedings of the National Academy of Sciences*, **20** (1934) 254–259. This quote appears on p. 258.
15. Interview of George Gamow by Charles Weiner on 25 April 1968, Niels Bohr Library & Archives, American Institute of Physics, College Park, MD USA, p. 57. www.aip.org/history-programs/niels-bohr-library/oral-histories/4325
16. Interview of George Gamow by Charles Weiner, pp. 96–97.

17. R.A. Alpher, H. Bethe, and G. Gamow, 'The Origin of Chemical Elements', *Physical Review*, **73** (1948) 803–804.

18. G. Gamow, letter to Hans Bethe, 15 April 1948. Quoted in Victor S. Alpher, 'Ralph A. Alpher, Robert C. Herman, and the Cosmic Microwave Background Radiation', 300–334. This quote appears on p. 307.

19. G. Gamow, 'The Evolution of the Universe', *Nature*, **162** (1948) 680–682; Ralph A. Alpher and Robert Herman, 'Evolution of the Universe', *Nature*, **162** (1948) 774–775.

20. Ralph A. Alpher and Robert Herman, *Genesis of the Big Bang*, Oxford University Press, 2001, p. 190.

21. R.A. Alpher and R.C. Herman, 'Remarks on the Evolution of the Expanding Universe', *Physical Review*, 75 (1949) 1089–1095.

22. Robert Herman, quoted in Victor S. Alpher, *Physics in Perspective*, **14** (2012), p. 310.

CHAPTER 5: PARAMETERS OF THE UNIVERSE

1. H. Bondi and T. Gold, 'The Steady-state Theory of the Expanding Universe', *Monthly Notices of the Royal Astronomical Society*, **108** (1948) 252–270.

2. F. Hoyle, 'A New Model for the Expanding Universe', *Monthly Notices of the Royal Astronomical Society*, **108** (1948) 372–382.

3. See Harry Nussbaumer and Lydia Bieri, *Discovering the Expanding Universe*, Cambridge University Press, 2009, p. 163.

4. Fred Hoyle, *The Nature of the Universe*, BBC Third Programme, 28 March 1949. A transcript of this broadcast was subsequently published in *The Listener* in April 1949. This quote is taken from Hoyle's original manuscript, selected pages of which are available to view online at http://www.joh.cam.ac.uk/library/special_collections/hoyle/exhibition/radio/.

5. R. Herman, quoted in Victor S. Alpher, *Physics in Perspective*, **14** (2012), p. 309.

6. J.S. Plaskett, *Popular Astronomy*, **47** (1939) 239–55.

7. W. Baade, 'The Period–Luminosity Relation of the Cepheids', *Publications of the Astronomical Society of the Pacific*, **68** (1956) 5–16: '... especially on cloudy winter nights', p. 8; '... and there was no a priori reason', p. 9.

8. See J. Francis Thackeray, 'Doubling the Age and Size of the Universe at the IAU in Rome in 1952: Contributions by David Thackeray, Walter Baade and Harlow Shapley', *South African Journal of Science*, **116** (2020) 1–2.

9. In his speech, Pius XII made no reference to Lemaître's cosmology. See (in French) https://www.vatican.va/content/pius-xii/fr/speeches/1952/documents/hf_p-xii_spe_19520907_la-presence.html.

10. P. Th. Oosterhof, ed., '28. Commission des Nebuleuses Extragalactiques', *Transactions of the IAU*, **8** (1954) 397–399.

11. A. Sandage, 'The First 50 Years at Palomar: 1949–1999. The Early Years of Stellar Evolution, Cosmology, and High-energy Astrophysics', *Annual Reviews of Astronomy and Astrophysics*, **37** (1999) 445–486.

12. Allan Sandage, quoted in Dennis Overbye, *Lonely Hearts of the Cosmos: The Story of the Scientific Quest for the Secret of the Universe*, Macmillan, London, 1991, pp. 28–29.

13. M.L. Humason, N.U. Mayall, and A.R. Sandage, 'Redshifts and Magnitudes of Extragalactic Nebulae', *Astronomical Journal*, **61** (1956) 97–162.

14. A. Sandage, 'Current Problems in the Extragalactic Distance Scale', *Astrophysical Journal*, **127** (1958) 513–526.
15. Allan Sandage, quoted in Dennis Overbye, p. 58.
16. D.E. Osterbrock, 'Rudolph Minkowksi 1895–1976', *National Academy of Sciences Biographical Memoirs*, **54** (1983) 280–281.
17. M. Schmidt, '3C 273: A Star-like Object with Large Redshift', *Nature*, **197** (1963) 1040.
18. H-Y. Chiu, 'Gravitational Collapse', *Physics Today*, **17** (1964) 21–34.
19. 'The Quasi-Quasars', *Time*, 18 June 1965.
20. Allan Sandage, Caltech press release quoted in Dennis Overbye, pp. 79–80.
21. Quoted in Dennis Overbye, p. 80.
22. Interview of Robert Dicke by Alan Lightman on 19 January 1988, Niels Bohr Library & Archives, American Institute of Physics, College Park, MD USA.
23. Quoted in Dennis Overbye, p. 130.
24. Interview of Jim Peebles by Martin Harwit on 27 September 1984, Niels Bohr Library & Archives, American Institute of Physics, College Park, MD USA.
25. Interview of Ralph Alpher and Robert Herman by Martin Harwit on 12 August 1983, Niels Bohr Library & Archives, American Institute of Physics, College Park, MD USA.
26. David Wilkinson, 'Measuring the Cosmic Microwave Background Radiation', in P. James E. Peebles, Lyman A. Page, Jr, and R. Bruce Partridge, eds, *Finding the Big Bang*, Cambridge University Press, 2009, p. 204. See also Peter G. Roll, p. 213.
27. P. James E. Peebles, Nobel Lecture, 8 December 2019, Nobelprize.org.
28. Arno Penzias, 'Detection at Bell Laboratories', in *Finding the Big Bang*, p. 148–151.
29. R.H. Dicke, P.J.E. Peebles, P.G. Roll, and D.T. Wilkinson, 'Cosmic Black-body Radiation', *Astrophysical Journal*, **142** (1965) 414–419.
30. Interview of Jim Peebles by Martin Harwit on 27 September 1984.
31. A.A. Penzias and R.W. Wilson, 'A Measurement of Excess Antenna Temperature at 4080 Mc/s', *Astrophysical Journal*, **142** (1965) 419–421.
32. Arno Penzias, 'Detection at Bell Laboratories', in *Finding the Big Bang*, p. 154.
33. 'Signals Imply a "Big Bang" Universe', *New York Times*, Friday 21 May 1965.

Chapter 6: Hubble Wars

1. R.B. Partridge, 'The Primeval Fireball Today', *American Scientist*, **57** (1969) 37–74.
2. A. Sandage and G.A. Tammann, 'A Composite Period–Luminosity Relation for Cepheids at Mean and Maximum Light', *Astrophysical Journal*, **151** (1968) 531–545.
3. Stephen Webb, *Measuring the Universe: The Cosmological Distance Ladder*, Springer-Verlag, Berlin, 1999, p. 179.
4. Quoted in Dennis Overbye, p. 169. Madore's paper was published in 1976: B.F. Madore, 'The Distance to NGC 2403', *Monthly Notices of the Royal Astronomical Society*, **177** (1976) 157–165.
5. Genco Guralp, 'Calibrating the Universe: The Beginning and the End of the Hubble Wars', in Oliver Schlaudt and Lara Huber, eds, *Standardization in Measurement: Philosophical, Historical and Sociological*, Pickering & Chatto, London, 2015.
6. V.C. Rubin, 'Differential Rotation of the Inner Metagalaxy', *Astronomical Journal*, **56** (1951) 47–48.

7. George Ogden Abell, 'The Distribution of Rich Clusters of Galaxies', Caltech Ph.D. thesis, 1957, p. 163.

8. Interview of Gérard de Vaucouleurs by Alan Lightman on 7 November 1988, Niels Bohr Library & Archives, American Institute of Physics, College Park, MD USA.

9. G. de Vaucouleurs, 'Further Evidence for a Local Super-cluster of Galaxies: Rotation and Expansion', *Astronomical Journal*, **63** (1958) 253–265.

10. Dennis Overbye, p. 267.

11. Lyman Spitzer Jr., 'Astronomical Advantages of an Extra-terrestrial Observatory', Project RAND, 30 July 1946, reproduced in John M. Logdson, ed., *Exploring the Unknown*, NASA History Office, Washington, DC, 1995, pp. 546–552.

12. Jimmy Lipp, quoted on the RAND Corporation website: https://www.rand.org/pubs/special_memoranda/SM11827.html.

13. Steven Weinberg, *The First Three Minutes: A Modern View of the Origin of the Universe*, Basic Books, New York, 1977, p. 72.

14. Robert H. Dicke, *Gravitation and the Universe: Jayne Lectures for 1969*, American Philosophical Society, Philadelphia, 1970, p. 62.

15. R.H. Dicke and P.J.E. Peebles, 'The Big Bang Cosmology—Enigmas and Nostrums' in S.W. Hawking and W. Israel, eds, *General Relativity: An Einstein Centenary Survey*, Cambridge University Press, 1979, pp. 504–517.

16. For a Friedmann-type universe with zero cosmological constant the relation is $\rho_c = 3H_0^2/8\pi G$, or $1.88 \times 10^{-33}H_0^2$ g/cm³ with H_0 in units of km/s/Mpc.

17. Lifshitz published a paper 'On the Gravitational Stability of the Expanding Universe' in the Soviet journal *Zhurnal Éksperimental'noĭ i Teoreticheskoĭ Fiziki* (Journal of Experimental and Theoretical Physics) in 1946. The English translation was republished in 2017 as a 'Golden Oldie' in the journal *General Relativity and Gravitation*, with an editorial note by George Ellis.

18. E.R. Harrison, 'Fluctuations at the Threshold of Classical Cosmology', *Physical Review D*, **1** (1970) 2726–2730; Y.B. Zeldovich, 'A Hypothesis, Unifying the Structure and Entropy of the Universe', *Monthly Notices of the Royal Astronomical Society*, **160** (1978) 1–3P.

19. J.P. Ostriker and P.J.E. Peebles, 'A Numerical Study of the Stability of Flattened Galaxies: or, Can Cold Galaxies Survive?', *Astrophysical Journal*, **186** (1973) 467–480.

20. J.P. Ostriker, P.J.E. Peebles, and A. Yahill, 'The Size and Mass of Galaxies, and the Mass of the Universe', *Astrophysical Journal*, **193** (1974) L1–L4.

21. Jeremiah P. Ostriker and Simon Mitton, *Heart of Darkness: Unravelling the Mysteries of the Invisible Universe*, Princeton University Press, 2013, p. 183.

22. H.W. Babcock, 'The Rotation of the Andromeda Nebula', *Lick Observatory Bulletin*, **19** (1939) 41–51 (No. 498).

23. A.B. Wyse and N.U. Mayall, 'Distribution of Mass in the Spiral Nebulae Messier 31 and Messier 33', *Astrophysical Journal*, **95** (1942) 24–47.

24. F. Zwicky, 'Die Rotverschiebung von Extragalaktischen Nebeln', *Helvetica Physica Acta*, **6** (1933) 110–127. This quote appears on p. 125. English translation using Google Translate.

25. Interview of Vera Rubin by Alan Lightman on 3 April 1989, Niels Bohr Library & Archives, American Institute of Physics, College Park, MD USA.

26. V.C. Rubin, W.K. Ford Jr, and N. Thonnard, 'Rotational Properties of 21 Sc Galaxies with a Large Range of Luminosities and Radii, from NGC 4605 ($R = 4$ kpc) to UGC 2885 ($R = 122$ kpc)', *Astrophysical Journal*, **238** (1980) 471–487. This quote appears on p. 485.

27. S.D.M. White and M.J. Rees, 'Core Condensation in Heavy Halos: A Two-stage Theory for Galaxy Formation and Clustering', *Monthly Notices of the Royal Astronomical Society*, **183** (1978) 341–385.

28. Quoted in Dennis Overbye, p. 273.

29. Allan Sandage and Gustav A. Tammann, 'Steps Towards the Hubble Constant. VII. Distances to NGC 2403, M101, and the Virgo Cluster Using 21 Centimeter Line Widths Compared with Optical Methods: The Global Value of H_0', *Astrophysical Journal*, **210** (1976) 7–24. This quote appears on p. 7.

30. R. Brent Tully and J. Richard Fisher, 'A New Method of Determining Distances to Galaxies', *Astronomy & Astrophysics*, **54** (1977) 661–673. This quote appears on p. 672.

31. John Huchra, 'The Hubble Constant', https://lweb.cfa.harvard.edu/~dfabricant/huchra/hubble/.

CHAPTER 7: THE INFLATIONARY UNIVERSE

1. I used Ned Wright's Javascript Cosmology Calculator to derive these look-back times: https://www.astro.ucla.edu/~wright/CosmoCalc.html.

2. R.K. Sachs and A.M. Wolfe, 'Perturbations of a Cosmological Model and Angular Variations of the Microwave Background', *Astrophysical Journal*, **147** (1967) 73–90.

3. J. Silk, 'Fluctuations in the Primordial Fireball', *Nature*, **215** (1967) 1155–1156.

4. This result is quoted in George Smoot and Keay Davidson, *Wrinkles in Time: Witness to the Birth of the Universe*, HarperCollins, New York, 1993, p. 118.

5. Smoot—Biographical. From Les Prix Nobel. The Nobel Prizes 2006, Editor Karl Grandin, [Nobel Foundation], Stockholm, 2007.

6. George Smoot and Keay Davidson, *Wrinkles in Time*, p. 109.

7. George F. Smoot—Biographical. From Les Prix Nobel.

8. G.F. Smoot, M.V. Gorenstein, and R.A. Muller, 'Detection of Anisotropy in the Cosmic Background Radiation', *Physical Review Letters*, **39** (1977) 898–901.

9. George Smoot and Keay Davidson, *Wrinkles in Time*, p. 138.

10. Alan H. Guth, *The Inflationary Universe: The Quest for a New Theory of Cosmic Origins*, Vintage, London, 1998, p. 176.

11. Guth, *The Inflationary Universe*, p. 179.

12. Paul J. Steinhardt and Neil Turok, *Endless Universe: Beyond the Big Bang*, Weidenfeld & Nicolson, London, 2007, p. 93.

13. Guth, *The Inflationary Universe*, p. 233.

14. Howard Georgi, interview with Robert Crease and Charles Mann, 29 January 1985. Quoted in Robert P. Crease and Charles C. Mann, *The Second Creation: Makers of the Revolution in Twentieth-century Physics*, Rutgers University Press, 1986, p. 400.

15. Richard P. Feynman, *What Do You Care What Other People Think? Further Adventures of a Curious Character*, Unwin Hyman, London, 1989, pp. 151, 153.

16. George Smoot and Keay Davidson, *Wrinkles in Time*, p. 216.

17. John Huchra, 'The Hubble Constant', https://lweb.cfa.harvard.edu/~dfabricant/huchra/hubble/.
18. Wendy Freedman, interview by David Zierler on 21 December 2020, Niels Bohr Library & Archives, American Institute of Physics, College Park, Maryland, USA.
19. M. Davis, J. Huchra, D.W. Latham, and J. Tonry, 'A Survey of Galaxy Redshifts II. The Large Scale Space Distribution', *Astrophysical Journal*, **253** (1982) 423–445. This quote appears on p. 423.
20. V. de Lapparent, M.J. Geller, and J.P. Huchra, 'A Slice of the Universe', *Astrophysical Journal*, **302** (1986) L1–L5. This quote appears on p. L1.
21. Margaret Geller, quoted in Jamie Murphy, 'Bubbles in the Universe', *Time*, 20 January 1986.
22. M.J. Geller and John P. Huchra, 'Mapping the Universe', *Science*, **246** (1989) 897–903. This quote appears on p. 900.
23. H.C. Arp, G. Burbidge, F. Hoyle, J.V. Narlikar, and N.C. Wickramasinghe, 'The Extragalactic Universe: An Alternative View', *Nature*, **346** (1990) 807–812. These quotes appear on pp. 807 and 812.

CHAPTER 8: DARK ENERGY AND THE ACCELERATING UNIVERSE

1. S. Hayakawa et al., 'Cosmological Implication of a New Measurement of the Submillimeter Background Radiation', *Publications of the Astronomical Society of Japan*, **39** (1987) 941–948; T. Matsumoto, S. Hayakawa, H. Matsuo, H. Murakami, S. Sato, A.E. Lange, and P.L. Richards, 'The Submillimeter Spectrum of the Cosmic Background Radiation', *Astrophysical Journal*, **329** (1988) 567–571.
2. George Smoot and Keay Davidson, *Wrinkles in Time*, p. 225.
3. J.C. Mather et al., 'A Preliminary Measurement of the Cosmic Microwave Background Radiation by the *Cosmic Background Explorer* (*COBE*) Satellite', *Astrophysical Journal*, **354** (1990) L37–L40.
4. Jeremiah P. Ostriker and Simon Mitton, *Heart of Darkness*, p. 165.
5. George Smoot and Keay Davidson, *Wrinkles in Time*, p. 282.
6. G.F. Smoot, et al., 'Structure in the COBE Differential Microwave Radiometer First-year Maps', *Astrophysical Journal*, **396** (1992) L1.
7. George Smoot and Keay Davidson, *Wrinkles in Time*, p. 289.
8. Joseph N. Tatarewicz, in Pamela E. Mack, ed., *From Engineering Science to Big Science: The NACA and NASA Collier Trophy Research Project Winners*, The NASA History Series, NASA, Washington, DC, 1998, p. 373.
9. Eric Chaisson, *The Hubble Wars: Astrophysics Meets Astropolitics in the Two-billion-dollar Struggle Over the Hubble Space Telescope*, Harper Collins, New York, 1994, p. 184.
10. H.C. Ford, et al. 'Narrowband HST Images of M87: Evidence for a Disk of Ionized Gas Around a Massive Black Hole', *Astrophysical Journal*, **435** (1994) L27–L30.
11. W. Freedman, interview by David Zierler on 21 December 2020, Niels Bohr Library & Archives, American Institute of Physics, College Park, Maryland, USA.
12. W.L. Freedman, et al. 'Distance to the Virgo Cluster Galaxy M100 from Hubble Space Telescope Observations of Cepheids', *Nature*, **371** (1994) 757–762.
13. Robert P. Kirshner, *The Extravagant Universe: Exploding Stars Dark Energy and the Accelerating Cosmos*, Princeton University Press, 2002, p. 168.

14. See Jim Baggott, *Atomic: The First War of Physics and the Secret History of the Atom Bomb: 1939–49*, Icon Books, London, 2009, pp. 162–7. Other alumni of the Los Alamos Ranch School included William Burroughs and Gore Vidal. According to Adam Riess (personal communication 13 August 2023), Colgate recalled seeing 'a guy in a pork pie hat' (Oppenheimer) and 'an army guy' (General Leslie Groves) inspect the school grounds.

15. Hans U. Nørgaard-Nielsen, Lief Hansen, Henning E. Jørgensen, Alfonso Aragón Salamanca, Richard S. Ellis, and Warrick J. Couch, 'The Discovery of a Type Ia Supernova at a Redshift of 0.31', *Nature*, **339** (1989) 523–5.

16. Saul Perlmutter, Nobel Lecture, 8 December 2011, p. 32.

17. Robert P. Kirshner, *The Extravagant Universe*, 2002, p. 185.

18. Richard Panek, *The 4% Universe: Dark Matter, Dark Energy, and the Race to Discover the Rest of Reality*, OneWorld Publications, London, 2011, p. 99.

19. A.G. Reiss, 'My Path to the Accelerating Universe', Nobel Lecture, 8 December 2011, p. 7.

20. Robert P. Kirshner, *The Extravagant Universe*, 2002, p. 184.

21. Riess, Nobel Lecture, 8 December 2011, p. 8.

22. Mario Hamuy, et al., 'A Hubble Diagram of Distant Type Ia Supernovae', *Astronomical Journal*, **109** (1995) 1–13.

23. A.G. Riess, W.H. Press, and R.P. Kirshner, 'Using Type Ia Supernova Light Curve Shapes to Measure the Hubble Constant', *Astrophysical Journal*, **438** (1995) L17–L20.

24. G. Lemaître, 'L' Expansion de l'Espace', *Revue des Questions Scientifiques* **17** (1931) 391–410. Quoted in English translation in Kragh, *Conceptions of Cosmos*. This quote appears on p. 154.

25. I used Ned Wright's Javascript Cosmology Calculator to derive these ages: https://www.astro.ucla.edu/~wright/CosmoCalc.html.

26. J.P. Ostriker and P.J. Steinhardt, 'The Observational Case for a Low-density Universe with a Non-zero Cosmological Constant', *Nature*, **377** (1995) 600–602.

27. L.M. Krauss and M.J. Turner, 'The Cosmological Constant is Back', *General Relativity and Gravitation*, **27** (1995) 1137–44.

28. Jeremiah P. Ostriker and Simon Mitton, *Heart of Darkness*, p. 223.

29. S. Perlmutter, et al., 'Measurements of the Cosmological Parameters Ω and Λ from the First Seven Supernovae at $z \geq 0.35$', *Astrophysical Journal*, **483** (1997) 565–581.

30. Gerson Goldhaber, 'The Acceleration of the Expansion of the Universe: A Brief Early History of the Supernova Cosmology Project (SCP)', *AIP Conference Proceedings*, **1166** (2009) 53; arXiv [Astro-ph.CO] 0907.3526v1, 21 July 2009, pp. 1–20.

31. Panek, *The 4% Universe*, 2011, p. 153.

32. Riess, Nobel Lecture, 8 December 2011, p. 13.

33. LBNL press release, 'Distant Exploding Stars Foretell Fate of the Universe', 8 January 1998. I am grateful to Adam Reiss for providing a copy of this document.

34. John Noble Wilford, 'New Data Suggest Universe Will Expand Forever', *New York Times*, 9 January 1998.

35. J. Glanz, 'Exploding Stars Point to a Universal Repulsive Force', *Science*, **279** (1998) 651–652.

36. Brian P. Schmidt, 'The Path to Measuring the Accelerating Universe', Nobel Lecture, 8 December 2011, p. 25.

37. Riess, Nobel Lecture, 8 December 2011, p. 16.

38. John Noble Wilford, 'Wary Astronomers Ponder an Accelerating Universe', *New York Times*, 3 March 1998.

39. A.G. Riess, et al., 'Observational Evidence from Supernovae for an Accelerating Universe and a Cosmological Constant', *Astronomical Journal*, **116** (1998) 1009–1038 (received 13 March 1998), and S. Perlmutter, et al., 'Measurements of Ω and Λ from 42 High-redshift Supernovae', *Astrophysical Journal*, **517** (1999) 565–586 (received 8 September 1998).

40. Panek, *The 4% Universe*, 2011, p. 240.

41. A.G. Riess, et al., 'The Farthest Known Supernova: Support for an Accelerating Universe and a Glimpse of the Epoch of Deceleration', *Astrophysical Journal*, **560** (2001) 49–71.

42. W.L. Freedman, et al., 'Final Results from the Hubble Space Telescope Key Project to Measure the Hubble Constant', *Astrophysical Journal*, **553** (2001) 47–72.

Chapter 9: Concordance

1. Martin White, 'Baryon Acoustic Oscillations: A Standard Ruler Method for Determining the Expansion Rate of the Universe', https://mwhite.berkeley.edu/BAO/bao_iucca.pdf.

2. D.J. Eisenstein, H-J. Seo, and M. White, 'On the Robustness of the Acoustic Scale in the Low-redshift Clustering of Matter', *Astrophysical Journal*, **664** (2007) 660–674.

3. R.A. Sunayev and Ya. B. Zeldovich, 'Small Scale Fluctuations of Relic Radiation', *Astrophysics and Space Science*, **7** (1970) 3–19, and P.J.E. Peebles and J.T. Yu, 'Primeval Adiabatic Perturbation in an Expanding Universe', *Astrophysical Journal*, **162** (1970) 815–836.

4. The length scale L is related to l according to $L = 2\pi c/H_0\Omega_M^{0.4}l$, where c is the speed of light and Ω_M is the density of matter. See Bernard J.T. Jones, Vincent J. Martinez, Enn Saar, and Virginia Trimble, 'Scaling Laws in the Distribution of Galaxies', *Reviews of Modern Physics*, **76** (2005) 1211–1266.

5. L. Bennett, et al., 'The Microwave Anisotropy Probe (MAP) Mission Concept', *Bulletin of the American Astronomical Society*, **127** (1995) 1385.

6. A. Lange, et al., 'The BOOMERANG Experiment', *Space Science Reviews*, **75** (1995) 145–150.

7. P. de Bernardis, et al., 'A Flat Universe from High-resolution Maps of the Cosmic Microwave Background Radiation', *Nature* **404** (2000) 955–959.

8. Andrew Lange, 'BOOMERANG—An Interview with Andrew Lange', ESA, 5 February 2001. https://www.esa.int/Science_Exploration/Space_Science/BOOMERANG_An_interview_with_Andrew_Lange.

9. C.B. Netterfield, et al., 'A measurement by BOOMERANG of Multiple Peaks in the Angular Power Spectrum of the Cosmic Microwave Background', *Astrophysical Journal*, **571** (2002) 604–614.

10. M. Tegmark, M. Zaldarriaga, and A.J.S. Hamilton, 'Latest Cosmological Constraints on the Densities of Hot and Cold Dark Matter' in D.B. Cline, ed., *Sources and Detection of Dark Matter and Dark Energy in the Universe*, Springer, Berlin (2001), and Max Tegmark, Matias Zaldarriaga, and Andrew J.S. Hamilton, 'Towards a Refined Cosmic Concordance Model: Joint 11-parameter Constraints From the Cosmic Microwave Background and Large-scale Structure', *Physical Review D*, **43** (2001) 043007.

11. J.C. Mather, L. Page, and P.J.E. Peebles, 'David Todd Wilkinson', *Physics Today*, **56** (2003) 76–77.

12. Wilkinson had a reputation among students as a showman. Will Kinney, now a physics professor at the University of Buffalo, SUNY, studied for his BA at Princeton, and took Wilkinson's freshman physics course in the autumn of 1982. He and his fellow students were terrified by the prospect of their first mid-term exam, in Princeton's Palmer Hall. Wilkinson watched as his Teaching Assistants (TAs) handed out the exam papers, and then they all promptly left the auditorium. This is Princeton's 'Honor Code': exams are unsupervised, and students are simply expected to behave honourably and with discretion. After just 15 minutes, a student from the back of the hall walked warily to the front, placed his completed exam paper on the table and made to leave. Another student cried out in disbelief: 'You're done? I'm only on question 1(b)!' Now furious, he stomped to the front and tore up the exam paper as the first student looked on, completely mortified. All the other students looked around nervously at each other. What should they do? The Honor Code meant nobody was in charge. Just then, the doors flew open and Wilkinson and his TAs rushed in, wearing full surgical gowns and masks, and without a word proceeded to carry the two students out of the hall. The doors closed. Everyone laughed. 'The tension gone, we finish the test', explains Kinney, 'To this day it is the only exam I have ever taken where people started to spontaneously chuckle in the middle of the test. A truly brilliant stunt'. And clearly very memorable.

13. C.L. Bennett, et al., 'First Year Wilkinson Microwave Anisotropy Probe (WMAP) Observations: Preliminary Maps and Basic Results', *Astrophysical Journal Supplement Series*, **148** (2003) 1–27, and G. Hinshaw, et al., 'First Year Wilkinson Microwave Anisotropy Probe (WMAP) Observations: The Angular Power Spectrum', *Astrophysical Journal Supplement Series*, **148** (2003) 135–159.

14. P.J.E. Peebles, *Cosmology's Century: An Inside History of Our Modern Understanding of the Universe*, Princeton University Press, 2020. This quote appears on p. 323.

Chapter 10: The Hubble Tension

1. S.A. Shectman, et al., 'The Las Campanas Redshift Survey', *Astrophysical Journal*, **470** (1996) 172–188.

2. Matthew Colless, 'The Great Cosmic Map', *Mercury*, March–April 2003, 30–36.

3. J. Peterson and G. Mackie, 'A Brief History of the Astrophysical Research Consortium and the Apache Point Observatory', *Journal of Astronomical History and Heritage*, **9** (2006) 109–118.

4. J.E. Gunn, et al., 'The 2.5 m Telescope of the Sloan Digital Sky Survey', *Astronomical Journal*, **131** (2006) 2332–2359.

5. Shaun Cole, 'Autobiography of Shaun Cole', https://www.shawprize.org/autobiography/shaun-cole/.

6. John A. Peacock, 'Autobiography of John A. Peacock', https://www.shawprize.org/autobiography/john-a-peacock/.

7. Ann K. Finkbeiner, *A Grand and Bold Thing: An Extraordinary New Map of the Universe Ushering in a New Era of Discovery*, Free Press, New York, 2010, p. 145.

8. J.R. Gott, et al., 'A Map of the Universe', *Astrophysical Journal*, **624** (2005) 463–484.

9. Colless, 'The Great Cosmic Map', p. 34.

10. S. Cole, et al., 'The 2dF Galaxy Redshift Survey: Power Spectrum Analysis of the Final Data Set and Cosmological Implications', *Monthly Notices of the Royal Astronomical Society*, **362** (2005) 505–534.

11. D.J. Eisenstein, et al., 'Detection of the Baryon Acoustic Peak in the Large-scale Correlation Function of SDSS Luminous Red Galaxies', *Astrophysical Journal*, **633** (2005) 560–574.

12. H-J. Seo and D.J. Eisenstein, 'Probing Dark Energy with Baryon Acoustic Oscillations From Future Large Galaxy Redshift Surveys', *Astrophysical Journal*, **598** (2003) 720–740.

13. Daniel Eisenstein, 'Autobiography of Daniel Eisenstein', https://www.shawprize.org/autobiography/daniel-eisenstein/.

14. W. Hu, 'Dark Energy Probes in Light of the CMB', in Sidney C. Wolff and Tod R. Lauer, eds, *Observing Dark Energy, Astronomical Society of the Pacific Conference Series*, **339** (2005) 215.

15. E. Komatsu, et al., 'Five-year *Wilkinson Microwave Anisotropy Probe* Observations: Cosmological Implications', *Astrophysical Journal Supplement Series*, **180** (2009) 330–376.

16. H.S. Leavitt, '1777 Variables in the Magellanic Clouds', *Annals of the Astronomical Observatory of Harvard College*, **60** (1908) 107.

17. A.G. Riess, et al., 'A Redetermination of the Hubble Constant with the *Hubble Space Telescope* from a Differential Distance Ladder', *Astrophysical Journal*, **699** (2009) 539–563.

18. Jan Tauber, https://sci.esa.int/web/planck/-/27639-map-blazes-the-way-for-planck-esas-probe-into-the-birth-of-the-universe.

19. E. Komatsu, et al., 'Seven-year *Wilkinson Microwave Anisotropy Probe* Observations: Cosmological Interpretation', *Astrophysical Journal Supplement Series*, **192** (2011) 1–47.

20. A.G. Riess, et al., 'A 3% Solution: Determination of the Hubble Constant with the *Hubble Space Telescope* and Wide Field Camera 3', *Astrophysical Journal*, **730** (2011) 1–18.

21. J. Dunkley, et al., 'The Atacama Cosmology Telescope: Cosmological Parameters from the 2008 Power Spectrum', *Astrophysical Journal*, **739** (2011) 1–20.

22. C.L. Bennett, et al., 'Nine-year *Wilkinson Microwave Anisotropy Probe (WMAP)* Observations: Final Maps and Results', *Astrophysical Journal Supplement Series*, **208** (2013) 1–54.

23. Planck Collaboration, '*Planck* 2013 Results. I. Overview of Products and Scientific Results', *Astronomy & Astrophysics*, **571** (2014) A1.

24. Jean-Loup Puget, 'Autobiography of Jean-Loup Puget', https://www.shawprize.org/autobiography/jean-loup-puget/.

25. Planck Collaboration, '*Planck* 2013 Results. XVI. Cosmological Parameters', *Astronomy & Astrophysics*, **571** (2014) A16.

26. Planck Collaboration, '*Planck* 2015 Results. XIII. Cosmological Parameters', *Astronomy & Astrophysics*, **594** (2016) A13; Planck Collaboration, '*Planck* 2018 Results. VI. Cosmological Parameters', *Astronomy & Astrophysics*, **641** (2020) A6.

27. P. James E. Peebles, Nobel Lecture, 8 December 2019, Nobelprize.org.

28. S. Alam, et al., 'The Clustering of Galaxies in the Completed SDSS-III Baryon Oscillation Spectroscopic Survey: Cosmological Analysis of the DR12 Galaxy Sample', *Monthly Notices of the Royal Astronomical Society*, **470** (2017) 2616–2652.

29. Simone Aiola, et al., 'The Atacama Cosmology Telescope: DR4 Maps and Cosmological Parameters', *Journal of Cosmology and Astroparticle Physics*, **2020** (2020) 047; Mathew S. Madhavacheril, et al., 'The Atacama Cosmology Telescope: DR6 Gravitational Lensing Map and Cosmological Parameters', arXiv [astro-ph.CO] 2304.05203v1, 11 April 2023, 1–30.

30. A.G. Riess, et al., 'A 2.4% Determination of the Local Value of the Hubble Constant', *Astrophysical Journal*, **826:56** (2016) 1–31.

31. A.G. Riess, et al., 'New Parallaxes of Galactic Cepheids from Spatially Scanning the *Hubble Space Telescope*: Implications for the Hubble Constant', *Astrophysical Journal*, **855:136** (2018) 1–18.

32. A.G. Riess, et al., 'Cosmic Distances Calibrated to 1% Precision with Gaia EDR3 Parallaxes and Hubble Space Telescope Photometry of 75 Milky Way Cepheids Confirm Tension With ΛCDM', *Astrophysical Journal Letters*, **908:L6** (2021) 1–11.

33. A.G. Riess, et al., 'A Comprehensive Measurement of the Local Value of the Hubble Constant With 1 kms^{-1}Mpc^{-1} Uncertainty from the Hubble Space Telescope and the SH$_0$ES Team', *Astrophysical Journal Letters*, **934:L7** (2022) 1–52.

34. A.G. Riess, 'The Expansion of the Universe is Faster Than Expected', *Nature Reviews Physics*, **2** (2020) 10–12.

35. C.D. Huang, et al., '*Hubble Space Telescope* Observations of Mira Variables in the SN Ia Host NGC 1559: An Alternative Candle to Measure the Hubble Constant', *Astrophysical Journal*, **889:5** (2020) 1–15.

36. J.P. Blakeslee, et al., 'The Hubble Constant from Infrared Surface Brightness Fluctuation Distances', *Astrophysical Journal*, **911:65** (2021) 1–12.

37. M.J. Reid, D.W. Pesce, and A.G. Riess, 'An Improved Distance to NGC 4258 and its Implications for the Hubble Constant', *Astrophysical Journal Letters*, **886:L27** (2019) 1–5.

38. D.W. Pesce, et al., 'The Megamaser Cosmology Project. XIII. Combined Hubble Constant Constraints', *Astrophysical Journal Letters*, **891:L1** (2020) 1–9.

39. The LIGO Scientific Collaboration and The Virgo Collaboration, The 1M2H Collaboration, The Dark Energy Camera GW-EM Collaboration and the DES Collaboration,. et al., 'A Gravitational-wave Standard Siren Measurement of the Hubble Constant', *Nature*, **551** (2017) 85–88.

40. V. Bonvin, et al., 'H$_0$LiCOW—V. New COSMOGRAIL Time Delays of HE 0435-1223. H_0 to 3.8% Precision from Strong Lensing in a Flat ΛCDM Model', *Monthly Notices of the Royal Astronomical Society*, **465** (2017) 4914–4930; Kenneth C. Wong, et al., 'H$_0$LiCOW—XIII. A 2.4 Per Cent Measurement of H_0 from Lensed Quasars: 5.3σ Tension Between Early- and Late-Universe Probes', *Monthly Notices of the Royal Astronomical Society*, **498** (2020) 1420–1439.

41. A.J. Shajib, et al., 'STRIDES: A 3.9 Per Cent Measurement of the Hubble Constant from the Strong Lens System DES J0408-5354', *Monthly Notices of the Royal Astronomical Society*, **494** (2020) 6072–6102.

42. In Sung Jang and Myung Gyoon Lee, 'The Tip of the Red Giant Branch Distances to Type Ia Supernova Host Galaxies. III. NGC 4038/39 and NGC 5584', *Astrophysical Journal*, **807:133** (2015) 1–12.

43. In Sung Jang and Myung Gyoon Lee, 'The Tip of the Red Giant Branch Distances to Type Ia Supernova Host Galaxies. V. NGC 3021, NGC 3370, and NGC 1309 and the Value of the Hubble Constant', *Astrophysical Journal*, **836:74** (2017) 1–13.

44. Wendy Freedman, interview by David Zierler on 21 December 2020, Niels Bohr Library & Archives, American Institute of Physics, College Park, Maryland, USA.

45. Wendy L. Freedman, et al., 'The Carnegie–Chicago Hubble Program. VIII. An Independent Determination of the Hubble Constant Based on the Tip of the Red Giant Branch', *Astrophysical Journal*, **882:34** (2019) 1–29; Wendy L. Freedman, et al., 'Calibration of the Tip of the Red Giant Branch', *Astrophysical Journal*, **891:57** (2020) 1–14; Wendy L. Freedman, 'Measurements of the Hubble Constant: Tensions in Perspective', *Astrophysical Journal*, **919:16** (2021) 1–22.

46. W.L. Freedman, 'Measurements of the Hubble Constant: Tensions in Perspective', *Astrophysical Journal*, **919:16** (2021) 1.

47. W. Yuan, et al., 'Consistent Calibration of the Tip of the Red Giant Branch in the Large Magellanic Cloud on the Hubble Space Telescope Photometric System and a Redetermination of the Hubble Constant', *Astrophysical Journal*, **886:61** (2019) 1–11.

48. Barry Madore, quoted in Natalie Wolchover, 'New Wrinkle Added to Cosmology's Hubble Crisis', *Quanta Magazine*, 26 February 2020.

49. K.A. Owens, W.L. Freedman, B.F. Madore, and A.J. Lee, 'Current Challenges in Cepheid Distance Calibrations Using Gaia Early Data Release 3', *Astrophysical Journal*, **927:8** (2022) 1–18.

50. R. Brent Tully, et al., 'The Extragalactic Distance Database', *Astronomical Journal*, **138** (2009) 323–31.

51. G.S. Anand, et al., 'Comparing Tip of the Red Giant Branch Distance Scales: An Independent Reduction of the Carnegie–Chicago Hubble Program and the Value of the Hubble Constant', *Astrophysical Journal*, **932:15** (2022) 1–18.

52. S.Dhawan, et al., 'A Uniform Type Ia Supernova Distance Ladder with the Zwicky Transient Facility: Absolute Calibration Based on the Tip of the Red Giant Branch Method', *Astrophysical Journal*, **934:185** (2022) 1–8.

53. D. Scolnic, et al., 'CATS: The Hubble Constant from Standardized TRGB and Type Ia Supernova Measurements', *Astrophysical Journal Letters*, **934:L31** (2023) 1–18.

EPILOGUE: DISCORDANCE

1. W.L. Freedman, 'Measurements of the Hubble Constant: Tensions in Perspective', *Astrophysical Journal*, **919:16** (2021) 1–22.

2. See, for example, Emma Chapman, *First Light: Switching on Stars at the Dawn of Time*, Bloomsbury Sigma, London, 2022.

3. B.E. Robertson, et al., 'Identification and Properties of Intense Star-forming Galaxies at z>10', *Nature Astronomy* Advanced Online Publication, April 2023; https://webbtelescope.org/contents/early-highlights/nasas-webb-reaches-new-milestone-in-quest-for-distant-galaxies.

4. P.G. Pérez-González, et al., 'A NIRCam-dark Galaxy Detected with the MIRI/F1000W Filter in the MIDIS/JADES Hubble Ultra Deep Field', *Astrophysical Journal Letters*, **969:L10** (2024) 1–12.

5. https://ceers.github.io/index.html.

6. Wendy L. Freedman, et al., 'Answering the Most Important Problem in Cosmology Today: Is the Tension in the Hubble Constant Real?', JWST Proposal Cycle 1, ID. #1995, March 2021.

7. W. Yuan, A.G. Riess, S. Casertano, and L.M. Macri, 'A First Look at Cepheids in a Type Ia Supernova Host with JWST', *Astrophysical Journal Letters*, **940:L17** (2022) 1–10.

8. A.G. Riess, et al., 'Crowded No More: The Accuracy of the Hubble Constant Tested with High Resolution Observations of Cepheids by *JWST*', arXiv [astro-ph.CO] 2307.01508v1, 28 July 2023, 1–20.

9. Wendy L. Freedman and Barry F. Madore, 'The Cepheid Extragalactic Distance Scale: Past, Present and Future', arXiv [astro-ph.CO] 2308.02474v1, 4 August 2023, 1–12.

10. Wendy L. Freedman, et al., 'Status Report on the Carnegie–Chicago Hubble Program (CCHP): Three Independent Astrophysical Determinations of the Hubble Constant Using the James Webb Space Telescope', arXiv [astro-ph.CO] 2408.06153v1, 12 August 2024, 1–61.

11. See, for example, M. Kamionkowski and A.G. Riess, 'The Hubble Tension and Early Dark Energy', *Annual Review of Nuclear and Particle Science*, **73** (2023) 153–80; arXiv [astro-ph.CO] 2211.04492v1, 8 November 2022.

12. Abdul Karim, M., et al., 'DESI DR2 Results II: Measurements of Baryon Acoustic Oscillations and Cosmological Constraints', arXiv [astro-ph.CO] 2503.14738v2, 26 March 2025.

13. E. Di Valentino, et al., 'In the Realm of the Hubble Tension—A Review of Solutions', *Classical and Quantum Gravity*, **38** (2021) 153001; N. Schöneberg, et al., 'The H_0 Olympics: A Fair Ranking of Proposed Models', *Physics Reports*, **984** (2022) 1–55.

14. S.S. McGaugh, 'A Tale of Two Paradigms: The Mutual Incommensurability of ΛCDM and MOND', *Canadian Journal of Physics*, **93** (2015) 250–9.

15. Halton Arp, et al., 'An Open Letter to the Scientific Community', *New Scientist*, 22 May 2004.

16. Anna Ijjas, Paul J. Steinhardt, and Abraham Loeb, 'Pop Goes the Universe', *Scientific American*, January 2017, pp. 32–39.

17. Paul J. Steinhardt, in G.W. Gibbons, S.W. Hawking and S.T.C. Siklos, eds, *The Very Early Universe, Proceedings of the Nuffield Workshop, Cambridge 21 June to 9 July, 1982*; Cambridge University Press, 1983.

18. A. Vilenkin, 'Birth of Inflationary Universes', *Physical Review D.*, **27** (1983) 2848; A.D. Linde, 'Eternally Existing Self-reproducing Chaotic Inflationary Universe', *Physics Letters B*, **175** (1986) 395–400.

19. John D. Barrow and Frank Tipler, *The Anthropic Cosmological Principle*, Oxford University Press, 1986.

20. Alan H. Guth, et al., 'A Cosmic Controversy', *Scientific American*, 10 May 2017: https:// www.scientificamerican.com/blog/observations/a-cosmic-controversy/.

21. Steven Weinberg, *Third Thoughts: The Universe We Still Don't Know*, Harvard University Press, 2018, p. 84.

22. See, for example, F. Melia, 'A Candid Assessment of Standard Cosmology', *Publications of the Astronomical Society of the Pacific*, **134** (2022) 121001. Melia's arguments are nicely summarized in Ethan R. Seigel, 'Eight Significant Shortcomings of the Standard Model of Cosmology', *New Ground*, **2** (2023) Article 11, which also includes a reply from the original author.

23. See, for example, Jim Baggott, *Quantum Space: Loop Quantum Gravity and the Search for the Structure of Space, Time, and the Universe*, Oxford University Press, 2018.

24. E.K. Anderson, et al., 'Observation of the Effect of Gravity on the Motion of Antimatter', *Nature*, **621** (2023) 716–22.

25. Albert Einstein, 'Induction and Deduction in Physics', *Berliner Tageblatt*, 25 December 1919; Alice Calaprice, ed., *The Ultimate Quotable Einstein*, Princeton University Press, 2011, p. 368.

26. https://www.pbs.org/newshour/show/how-the-trump-administrations-plans-to-slash-nasas-budget-will-impact-science.

Bibliography

Alpher, Ralph A., and Robert Herman, *Genesis of the Big Bang*, Oxford University Press, 2001.

Baggott, Jim, *Higgs: The Invention and Discovery of the 'God Particle'*, Oxford University Press, 2012.

Baggott, Jim, *Quantum Space: Loop Quantum Gravity and the Search for the Structure of Space, Time, and the Universe*, Oxford University Press, 2018.

Barrow, John D., and Frank Tipler, *The Anthropic Cosmological Principle*, Oxford University Press, 1986.

Bertone, Gianfranco, *A Tale of Two Infinities: Gravitational Waves and the Quantum Origin of the Universe's Biggest Mysteries*, Oxford University Press, 2021.

Calaprice, Alice, ed., *The Ultimate Quotable Einstein*, Princeton University Press, 2011.

Chaisson, Eric, *The Hubble Wars: Astrophysics Meets Astropolitics in the Two-billion-dollar Struggle Over the Hubble Space Telescope*, HarperCollins, New York, 1994.

Christianson, Gale E., *Edwin Hubble: Mariner of the Nebulae*, Institute of Physics Publishing, Bristol, 1997.

Crease, Robert P., and Charles C. Mann, *The Second Creation: Makers of the Revolution in Twentieth-century Physics*, Rutgers University Press, 1986.

Eddington, A.S., *Stellar Movements and the Structure of the Universe*, MacMillan & Co., London, 1914.

Eddington, A.S., *The Mathematical Theory of Relativity*, Cambridge University Press, 1923.

Eddington, A.S., *The Expanding Universe*, Cambridge University Press, 1933.

Einstein, Albert, *Relativity: The Special and the General Theory*, 100th Anniversary Edition, Princeton University Press, 2015.

Farmelo, Graham, ed., *It Must be Beautiful: Great Equations of Modern Science*, Granta Books, London, 2002.

Ferris, Timothy, *The Red Limit: The Search for the Edge of the Universe*, William Morrow & Co., New York, 1977.

Feynman, Richard P., *What Do You Care What Other People Think? Further Adventures of a Curious Character*, Unwin Hyman, London, 1989.

Friedmann, Alexander A., *The World as Space and Time*, translated by Svetla Kirilova-Petkova and Vesselin Petkov, edited by Vesselin Petkov, Minkowski Institute Press, Montreal, 2014.

Gamow, George, *My World Line: An Informal Autobiography*, Viking Press, New York, 1970.

Guth, Alan H., *The Inflationary Universe: The Quest for a New Theory of Cosmic Origins*, Vintage, London, 1998.

Hawking, Stephen, *A Brief History of Time: From the Big Bang to Black Holes*, Bantam Press, 1988.

Hawking, S.W., and W. Israel, eds, *General Relativity: An Einstein Centenary Survey*, Cambridge University Press, 1979.

Hubble, Edwin, *The Realm of the Nebulae*, Yale University Press, 1936.

Isaacson, Walter, *Einstein: His Life and Universe*, Simon & Shuster, New York, 2007.

Johnson, George, *Miss Leavitt's Stars*, W.W. Norton & Company, New York, 2005.

Jones, Bernard J.T., *Precision Cosmology: The First Half Million Years*, Cambridge University Press, 2017.

Jones, Bessie Saban, and Lyle Gifford Boyd, *The Harvard College Observatory: The First Four Directorships 1839–1919*, Belknap Press, Cambridge, MA, 1971.

Kennedy, J.B., *Space, Time and Einstein: An Introduction*, Acumen, Chesham, 2003.

Kennedy, Robert E., *A Student's Guide to Einstein's Major Papers*, Oxford University Press, 2012.

Kirshner, Robert P., *The Extravagant Universe: Exploding Stars, Dark Energy and the Accelerating Cosmos*, Princeton University Press, 2002.

Kragh, Helge S., *Conceptions of Cosmos*, Oxford University Press, 2007.

Krauss, Lawrence M., *A Universe from Nothing: Why There is Something Rather than Nothing*, Simon & Schuster, London, 2012.

Mach, Ernst, *The Science of Mechanics*, Open Court Press, Chicago, IL, 1893.

Mo, Hojun, Frank van den Bosch, and Simon White, *Galaxy Formation and Evolution*, Cambridge University Press, 2010.

Nussbaumer, Harry, and Lydia Bieri, *Discovering the Expanding Universe*, Cambridge University Press, 2009.

Ostriker, Jeremiah P., and Simon Mitton, *Heart of Darkness: Unravelling the Mysteries of the Invisible Universe*, Princeton University Press, 2013.

Overbye, Dennis, *Lonely Hearts of the Cosmos: The Story of the Scientific Quest for the Secret of the Universe*, Macmillan, London, 1991.

Pais, Abraham, *Subtle is the Lord: The Science and the Life of Albert Einstein*, Oxford University Press, 1982.

Panek, Richard, *The 4% Universe: Dark Matter, Dark Energy, and the Race to Discover the Rest of Reality*, OneWorld Publications, London, 2011.

Peebles, P. James E., The Large-scale Structure of the Universe, Princeton University Press, 1980.

Peebles, P. James E., *Cosmology's Century: An Inside History of Our Modern Understanding of the Universe*, Princeton University Press, 2020.

Peebles, P. James E., Lyman A. Page, Jr, and R. Bruce Partridge, eds, *Finding the Big Bang*, Cambridge University Press, 2009.

Rees, Martin, *Just Six Numbers: The Deep Forces that Shape the Universe*, Phoenix, London, 2000.

Rovelli, Carlo, *Seven Brief Lessons on Physics*, Allen Lane, London, 2015.

Ryan, Sean G., and Andrew J. Norton, *Stellar Evolution and Nucleosynthesis*, Cambridge University Press, 2010.

Shapley, Harlow, *Through Rugged Ways to the Stars: The Reminiscences of an Astronomer*, Charles Scribner's Sons, New York, 1969.

Singh, Simon, *Big Bang: The Most Important Scientific Discovery of All Time and Why You Need to Know About It*, Harper Perennial, London, 2005.

Smoot, George, and Keay Davidson, *Wrinkles in Time: Witness to the Birth of the Universe*, HarperCollins, New York, 1993.

Sobel, Dava, *The Glass Universe: The Hidden History of the Women Who Took the Measure of the Stars*, Fourth Estate, London, 2016.

Steinhardt, Paul J., and Neil Turok, *Endless Universe: Beyond the Big Bang*, Weidenfeld & Nicolson, London, 2007.

Voller, Ron, *Hubble, Humason, and the Big Bang: The Race to Uncover the Expanding Universe*, Springer Praxis, Cham, 2021.

Waller, William H., *The Milky Way: An Insider's Guide*, Princeton University Press, 2013.

Webb, Stephen, *Measuring the Universe: The Cosmological Distance Ladder*, Springer-Verlag, Berlin, 1999.

Weinberg, Steven, *The First Three Minutes: A Modern View of the Origin of the Universe*, Basic Books, New York, 1977.

Weinberg, Steven, *Cosmology*, Oxford University Press, 2008.

Weinberg, Steven, *Third Thoughts: The Universe We Still Don't Know*, Harvard University Press, 2018.

Wheeler, John Archibald, with Kenneth Ford, *Geons, Black Holes and Quantum Foam: A Life in Physics*, W.W. Norton & Company, New York, 1998.

Index

Note: figures and tables are indicated by italic *f* and *t* following the page number. Greek characters are filed according to their English equivalents, e.g. 'Λ' is filed as 'lambda'.

For the benefit of digital users, indexed terms that span two pages (e.g., 52–53) may, on occasion, appear on only one of those pages.